About IFPRI

The International Food Policy Research Institute (IFPRI), established in 1975, provides research-based policy solutions to sustainably reduce poverty and end hunger and malnutrition. The Institute conducts research, communicates results, optimizes partnerships, and builds capacity to ensure sustainable food production, promote healthy food systems, improve markets and trade, transform agriculture, build resilience, and strengthen institutions and governance. Gender is considered in all of the Institute's work. IFPRI collaborates with partners around the world, including development implementers, public institutions, the private sector, and farmers' organizations. IFPRI is a member of the CGIAR Consortium.

About IFPRI's Peer Review Process

IFPRI books are policy-relevant publications based on original and innovative research conducted at IFPRI. All manuscripts submitted for publication as IFPRI books undergo an extensive review procedure that is managed by IFPRI's Publications Review Committee (PRC). Upon submission to the PRC, the manuscript is reviewed by a PRC member. Once the manuscript is considered ready for external review, the PRC submits it to at least two external reviewers who are chosen for their familiarity with the subject matter and the country setting. Upon receipt of these blind external peer reviews, the PRC provides the author with an editorial decision and, when necessary, instructions for revision based on the external reviews. The PRC reassesses the revised manuscript and makes a recommendation regarding publication to the director general of IFPRI. With the director general's approval, the manuscript enters the editorial and production phase to become an IFPRI book.

Socioeconomic Considerations in Biosafety Decisionmaking

Methods and Implementation

Edited by Daniela Horna, Patricia Zambrano, and José Falck-Zepeda

Facilitated by **IFPRI**

A Peer Reviewed Publication

International Food Policy Research Institute
Washington, DC

International Food Policy Research Institute
2033 K Street, NW
Washington, DC 20006-1002, USA
Telephone: +1-202-862-5600
www.ifpri.org

DOI: http://dx.doi.org/10.2499/9780896292079

Library of Congress Cataloging-in-Publication Data

Socioeconomic considerations in biosafety decisionmaking : methods and
   implementation / edited by Daniela Horna, Patricia Zambrano, and
   José Falck-Zepeda.
       p.   cm.
Includes bibliographical references and index.
ISBN 0-89629-207-X (alk. paper)
    1. Cotton—Biotechnology—Uganda.   2. Cotton—Biotechnology—
Economic aspects—Uganda.   3. Cotton—Biotechnology—Uganda—
Safety measures.   4. Transgenic plants—Risk assessment.   I. Horna
Rodríguez, Julia Daniela.   II. Zambrano, Patricia.   III. Falck-Zepeda,
José Benjamin.   IV. International Food Policy Research Institute.
SB249.S68   2013
338.1'7351—dc23   2013021132

Contents

Tables and Figures

Tables

Figures

Abbreviations and Acronyms

BPA	Bukalasa Pedigree Albar
Bt	*Bacillus thuringiensis* [insect resistant]
Bt/RR	Bt and Roundup Ready, stacked gene
CDO	Cotton Development Organisation
CFT	confined field trials
CGE	computable general equilibrium
CPB	Cartagena Protocol on Biosafety
FAO	Food and Agriculture Organization of the United Nations
GM	genetically modified
GTAP	Global Trade Analysis Project
HT	herbicide tolerant
IFOAM	International Federation of Organic Agricultural Movements
IR	insect resistant
IRR	internal rate of return
NARO	National Agricultural Research Organisation
NBC	National Biosafety Committee
NOGAMU	National Organic Agricultural Movement of Uganda
PBS	Program for Biosafety Systems
R&D	research and development

UNAS	Uganda National Academy of Science
UNCST	Uganda National Council for Science and Technology
UNHS	Uganda National Household Survey
USAID	US Agency for International Development

Foreword

The expanded use of genetically modified (GM) crops in the developing world has opened a debate about the inclusion of socioeconomic considerations in the biosafety regulatory process through which these crops are approved. In the recent past, the apparent consensus and common practice was to exclude such considerations from the regulatory process, a position backed by environmental risk assessors and many regulatory experts. Despite this position, active civil society organizations and some social scientists have been successful in pushing for the inclusion of socioeconomic considerations. Currently, several developing countries, including South Africa, Indonesia, and other country parties to the Cartagena Protocol on Biosafety, are exploring the option of including such considerations in their decisionmaking process. How to define the analysis and scope of socioeconomic considerations and insert them into a functional biosafety system are questions still awaiting answers.

A robust technology-assessment process that includes socioeconomic considerations must address the following questions: At what stage or stages of the regulatory process should socioeconomic considerations be included? What would be the adequate level and scope of analysis for these considerations? How are they going to be included in the decisionmaking process? How will the results of an assessment be judged in relation to biophysical evaluations?

In addressing these questions, the authors have drawn from their cumulative experience in several countries in Africa, Asia, and Latin America, where they have conducted ex ante and ex post impact evaluations of GM technologies. *Socioeconomic Considerations in Biosafety Decisionmaking: Methods and Implementation* serves as a valuable and timely guide for government research staff, academics, independent consultants, and others responsible for implementing the socioeconomic assessment of GM technologies as part of the biosafety approval process. Using the case of GM cotton in Uganda, the authors propose and develop a methodological framework for the inclusion of socioeconomic considerations in biosafety evaluations.

Shenggen Fan
Director General, IFPRI

Acknowledgments

This study was made possible through the support of the Program for Biosafety Systems (PBS). PBS is funded by the US Agency for International Development (USAID). The opinions expressed herein are those of the authors and do not necessarily reflect the views of USAID or its missions worldwide.

Representatives of public and international institutions in Uganda kindly met with us and shared their knowledge about cotton production in Uganda. We are particularly grateful to Arthur Makara from the National Biosafety Committee, Muzoora Hans Winsor and Damalie Lubwana from the Cotton Development Organisation, Amos Tindyebwa from the Uganda Export Promotion Board, Sunday Godfrey from the National Agricultural Research Organisation, and Herbert Kirunda from the Agricultural Productivity Enhancement Program.

We thank Eva Schiffer for guiding and facilitating the use of Net-Map in the analysis of the institutional setting. We are particularly grateful to Marnus Gouse for reviewing the study and providing valuable comments. We also extend our thanks to John Baffes and Laoura Maratou for their feedback on earlier versions of this work.

This work would not have been possible without the participation of cotton producers in Lira and Kasese, who generously agreed to participate in our household survey and shared their experiences.

Chapter 1

Introduction

Daniela Horna, Patricia Zambrano, and José Falck-Zepeda

I n a world afflicted by poverty, food insecurity, species loss, and ecosystem destruction, the question of how to improve livelihoods and at the same time fulfill international environmental commitments poses a tremendous challenge (Young 2004). Although it is true that ending poverty and ensuring food security are complex goals that require global multilateral actions, the implementation of specific strategies, such as the introduction of genetically modified (GM) crops, has the potential to contribute to achieving these goals. Even though the number of countries adopting GM crops has been expanding over the past 16 years, growing from a handful to more than 29 developed and developing countries worldwide, there are many more countries still working on the establishment of the regulatory frameworks necessary for the assessment and commercial approval of GM crops. This study provides a methodological framework that can be adapted and adopted to support the regulatory process for GM crops in the increasingly common cases of countries that are opting to include socioeconomic considerations as part of their biosafety regulation.

The assessment and approval of GM crops includes the biosafety regulatory process as agreed by signatories to the Cartagena Protocol on Biosafety (CPB).[1] A responsible introduction of GM crops should follow the biosafety regulations mandated by each country, which must include a risk assessment prior to approval. Article 26.1 of the CPB states that the inclusion of socioeconomic considerations as part of this risk assessment is not mandatory, thus leaving entirely to the countries the decision of whether to include such considerations. Many countries have in fact opted to include, or are considering including, socioeconomic considerations in their decisionmaking processes. Some of these countries have even expanded the limited scope of Article 26.1 to include not only socioeconomic considerations but also ethical,

---

1 Although there is no agreed definition of biosafety in the CPB or elsewhere, here we use the term biosafety as in the CPB to refer to all critical safety aspects related to the transboundary movement, transit, handling, and use of living modified organisms (LMOs) that may have an adverse effect on the conservation and sustainable use of biological diversity.

religious, and aesthetic issues. Given the difficulty of measuring these types of considerations, the complexity of the assessment and decisionmaking process has and will likely continue to be greatly increased.

The inclusion of socioeconomic considerations in biosafety regulations, specifically in the approval process for GM crops, is a topic of discussion in both developed and developing economies. Many African countries have already decided to include these considerations in their regulatory processes, even though no clear guidelines exist as to how these considerations will be evaluated. The specific objective of this study is to provide guidance on how to conduct an ex ante economic assessment of a GM crop when such an assessment becomes part of the crop's approval process. We use the case of GM cotton in Uganda to illustrate this process, as the knowledge generated by this evaluation can be used for other biosafety regulatory and technology decisionmaking processes. Uganda is still developing official policies and regulations with regard to biotechnology and biosafety. If the government decides to include socioeconomic considerations in these regulations, it will be crucial to implement effective strategies for impact assessments (Kikulwe 2010).

The present evaluation was implemented at the request of the National Biosafety Committee of Uganda (NBC), the competent regulatory authority in the country.[2] The growing demand for further guidance on socioeconomic assessments became the rationale for this monograph. This demand is likely to further expand as deliberations under the CPB and at the national level provide further support for some countries to include socioeconomics in their decisionmaking.

A point worth highlighting is that a technology assessment in the context of a biosafety regulatory process differs from an ordinary technology assessment in two main ways. First, the selection of methods to be included in any assessment framework is limited by the need to adjust to a regulatory process. The longer it takes to carry out a technology assessment, the higher the costs

---

[2] During the biosafety assessment process for a GM cotton technology submitted by the National Agricultural Research Organisation of Uganda (NARO) to the NBC, some stakeholders raised the issues of potential socioeconomic and institutional constraints and the impacts from adopting such a technology. The NBC requested the Program for Biosafety Systems (PBS), a program facilitated by the International Food Policy Research Institute, to undertake such an assessment as may be required in the future for biosafety regulatory approval in Uganda (although not currently included in national regulations). PBS undertook the development and deployment of a relatively small field study with a limited budget, which would satisfy the regulatory authority. Although this socioeconomic assessment was in response to a request made by the competent authority in Uganda to meet a regulatory requirement, the PBS team envisions that the binding methodological and research implementation challenges that both the research and regulatory and policymaking communities face in Uganda could be extended to other developing countries considering the approval of a GM crop technology.

become for the regulatory process. Moreover, methods that require intensive data collection would have implications for the time and resources used to implement them and therefore for the total regulatory costs. Thus, the challenge is to use methods and tools that are scientifically sound and can help with decisionmaking without adding unnecessary costs to compliance with the development and regulatory processes. The latter is particularly critical for financially constrained public organizations in developing countries that are moving their projects through the regulatory process.

Second, technology assessments that contribute to the approval or rejection of a specific GM crop technology will need to be ex ante. Ex ante assessments require certain methods to collect data and generate information, and they involve limited availability of data and information. The effect of the technology in the field can only be assumed, and it is on the reliability of this assumption that the relevance and applicability of the assessment rest. The only situation where a socioeconomic assessment will consider ex post evaluations is when the regulatory authorities consider post-release monitoring, usually as a condition for temporary permits for commercial use.

To ensure a well-informed public and sound decisionmaking related to GM crops, it is essential to understand as comprehensively as possible the benefits and risks faced by the different actors involved in the crop value chain and the contribution of this technology to agricultural growth. From a decisionmaking point of view, it is also essential to understand the limitations of implementing such evaluations, given the time and budget constraints of approval through a regulatory process. This study uses a narrow definition of socioeconomic considerations to start the discussion of the fundamental requirements for a biosafety regulation. The study does not evaluate impacts on health and biodiversity or discuss property rights. It focuses on other key socioeconomic considerations, such as impacts on farmers, the national economy, and trade. In future evaluations of GM cotton, it would be desirable to include analysis of such topics as intellectual property rights, impacts on traditional knowledge, and environmental impacts. Nevertheless, such inclusion may be limited ex ante by data constraints, as observed data can only become available after adoption of the technologies under approval.

The current study uses a methodological framework (Falck-Zepeda et al. 2006; Horna et al. 2008; Smale et al. 2009) that has been modified to take into account the challenges noted above in implementing an assessment in the context of a regulatory approval process. Unlike in previous studies, we first evaluate how the current conditions, including limiting factors and institutional capacity, affect adoption and delivery of the technology. From this initial assessment, we build assumptions that support the analytical models and data collection activities that lead to the assessment of the economic impact of GM cotton on farms, industry, and trade. We propose this approach as a standard operating procedure for future socioeconomic assessments of GM crop technologies for the purposes of biosafety regulatory approval. We expect that countries currently seeking guidance on how to include socioeconomic considerations will benefit from this proposed approach: it helps define the process and creates inroads to discussing critical issues, such as cost and time efficiencies, regulatory impacts assessments, assessment timing and approach, and the definition of decisionmaking standards.

Adoption and Impact of GM Crop Technologies

The adoption of GM varieties worldwide has expanded considerably over the years. When GM crops were first commercialized in 1996, only four countries planted transgenic crops, covering an area of just more than 11 million hectares. By 2011, however, this area had expanded to 160 million hectares in 29 countries (Figure 1.1). Soybeans, maize, cotton, and canola continue to dominate among GM crops, although there are others now in the field.

FIGURE 1.1   Worldwide area devoted to genetically modified (GM) crops and number of adopting countries, 1997–2011

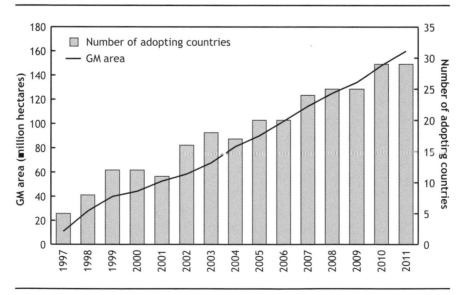

Sources: Based on data from James (2007, 2008, 2009, 2010, 2011).

The United States holds more than 40 percent of the world's total GM area, although there are 18 developing countries that have embraced this technology. Herbicide-tolerant (HT) soybean has been the most widespread GM crop, followed by insect-resistant (IR) and HT maize and cotton. In the case of cotton, GM varieties have been commercialized in 12 countries. The first countries to plant insect-resistant (*Bacillus thuringiensis,* or Bt) cotton were Mexico and the United States in 1996, followed by Argentina, Australia, and South Africa in 1998. China followed in 2000, and Colombia, India, and Brazil two years later. The latest adopters have been Burkina Faso in 2008, Myanmar and Pakistan in 2010, and Sudan in 2012 (James 2007, 2008, 2009, 2010, 2011, 2012). In parallel, the areas of non-GM conventional cotton have been decreasing in favor of IR and HT cotton (Figure 1.2). James (2011) estimates that more than 68 percent of world cotton is now planted to some type of GM seed, including both varieties and hybrids.

At first glance, the literature on the worldwide impact of GM crop adoption can appear to be both extensive and contradictory, reflecting the active and opposing positions that have characterized the debate since these crops were first released. The impact on developing economies is equally contradictory. In an effort to assess the benefits for poor farmers from the adoption

FIGURE 1.2   Worldwide areas devoted to genetically modified and conventional cotton (million hectares), 2000–2009

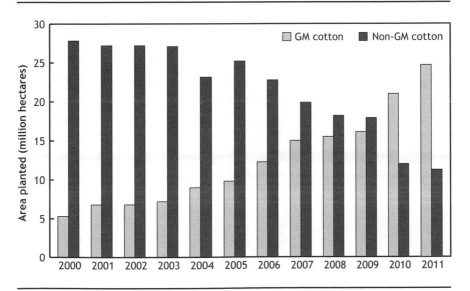

Sources: Based on James (2007, 2008, 2009).

of GM crops, the International Food Policy Research Institute published in 2009 a review of the economic impact of these crops on developing economies and the methods used to evaluate this impact (Smale et al. 2009). The study focused on the peer-reviewed applied economics literature. Of the 153 articles reviewed, 63 were on IR cotton, most of them studies from China, India, and South Africa. The review reveals that the impact on farmers was studied most often, but the reviewed articles also evaluated the impact on consumers, international trade, and the cotton industry overall.

The same study documented the performance of GM cotton. Table 1.1 presents the parameters from reported field surveys (as summarized by Smale, Niane, and Zambrano 2010). The use of Bt cottonseed on average increases yields, reduces the use of insecticides, and increases profitability, with better comparative results for some countries. However, the degree of variability among and within countries, and in some cases even within specific case studies, is rather high. Overall, results at the farm level are not homogenous and cannot be generalized across regions, as there is considerable vari-ation. Farm-level results are tied to specific sites and agroecological areas and therefore to climatic conditions and pest pressures that can vary from one year to another. Thus, as observed in Table 1.1, there are documented cases where Bt cotton has resulted in increased use of insecticides and lowered yields and profits (Smale, Niane, and Zambrano 2010). The conclusion of the IFPRI review, nevertheless, is that Bt cotton "on average ... [does] provide economic advantages for adopting farmers" (Smale et al. 2009, 32). The authors also note that, among all GM crops reviewed, the most successful case appears to be cotton.

Other authors have also reviewed the literature on the impact of GM crops (Raney 2006; Qaim 2009; National Research Council 2010), but they have concentrated more on the findings than the actual methods. Qaim (2009) reported that GM crops were beneficial to farmers and consumers and produced large aggregate welfare gains, with farmers in developing countries benefiting more than farmers in developed countries. Raney (2006) concluded that the poorest farmers in developing countries could benefit from transgenic crops but that ensuring their access to GM technology remained a formidable challenge.

Sexton and Zilberman (2011) made the first econometric estimation of the global effects of GM crops on yield increase using data from the Food and Agriculture Organization of the United Nations (FAO) for 100 countries and eight crops and adoption rates from the International Service for the Acquisition of Agri-biotech Applications. This estimation shows that GM crops produce an increase in yields relative to non-GM ones that is statistically

TABLE 1.1 Comparison of the performance of Bt and non-Bt cotton varieties/hybrids, by parameter and country, 1997–2007

Parameter	India	China	South Africa	Argentina and Mexico
Yield advantage (number of observations)	40	11	41	9
Min (percent)	−17	−6	−36	−3
Median (percent)	42	6	56	32
Max (percent)	92	55	129	65
Insecticide cost (number of observations)	29	7	29	8
Min (percent)	−83	−82	−95	−81
Median (percent)	−3	−66	−53	−51
Max (percent)	83	−56	68	−2
Profit (number of observations)	16	n.a.	n.a.	n.a.
Min (percent)	−65	n.a.	n.a.	n.a.
Median (percent)	47	n.a.	n.a.	n.a.
Max (percent)	136	n.a.	n.a.	n.a.

Source: Smale, Niane, and Zambrano (2010).
Notes: There can be more than one observation per study. Bt = insect resistant; max = maximum performance of Bt cotton relative to non-Bt cotton; median = median performance of Bt cotton relative to non-Bt cotton; min = minimum performance of Bt cotton relative to non-Bt cotton; n.a. = not available.

and economically significant. GM cotton's estimated yield increase is an impressive average of 65 percent and is bigger for developing countries than for industrialized ones.

The most recent efforts to assess on-farm benefits of GM crops are a couple of meta-analyses done by Finger et al. (2011) and Areal, Riesgo, and Rodriguez-Cerezo (2012). The first study analyzed Bt and conventional cotton performance data collected from 58 country reports or studies, mainly from Australia, China, India, South Africa, and the United States. The authors found that Bt cotton yield is about 46 percent higher than conventional cotton yield; however, this result is statistically significant only for India. Moreover, gross profit margins for Bt cotton are significantly higher (84 percent), despite the higher seed cost of Bt cotton relative to conventional cotton. Although not statistically significant, the results also show lower pesticide costs and management costs for Bt cotton. Finger et al. (2011) therefore suggest caution in the interpretation of these data, as the number of observations from India is large and there is great variability among and within countries. These conclusions are quite similar to those obtained by Smale, Niane, and Zambrano (2010).

Areal, Riesgo, and Rodriguez-Cerezo (2012) reviewed the results of 56 scientific articles, the majority of them from developing countries, particularly India (19) and South Africa (12). Their overall assessment is that the data evaluated confirm that GM crop yields outperform those of conventional crops, although this assessment can only be generalized for Bt cotton and not necessarily for HT cotton varieties. A very interesting result of this global meta-analysis is that the gains observed both in yields and gross margins as a result of using GM rather than conventional crops are higher for developing countries than for developed countries. The authors add a note of caution regarding endogeneity, as the data do not allow discerning what the effects of the technology itself are and what is explained by farmers' skills.

Socioeconomic Assessments in the Biosafety Regulatory Process

The development of GM crop technologies in the late 1970s and early 1980s compelled the design and implementation of protocols and procedures to regulate manipulation and ensure that safety assessments were undertaken for these technologies. Biosafety regulatory design was originally based on existing regulatory systems in agriculture and other sectors of the economy. As explained by Falck-Zepeda and Zambrano (2011), these early experiences were later incorporated into the 2000 CPB to the Convention on Biological Diversity.[3] The CPB is primarily an international agreement that makes biosafety assessments a precondition for approving GM crops for transboundary trade (Falck-Zepeda and Zambrano 2011). The protocol became operational in 2003, and since then it has become a driving force for the development of national regulatory systems.

Currently, however, there is no clear rule, even for CPB parties, about including socioeconomic considerations in the biosafety regulatory process for approval of GM technologies. The CPB seeks to protect biological diversity from the potential risks posed by "living modified organisms (LMOs) resulting from modern biotechnology" (CPB 2000). The protocol proposes the possibility of including socioeconomic considerations in biosafety regulatory approval processes and decisionmaking for GM crops. Article 26.1 of the protocol states:

---

3   Information about the Convention on Biological Diversity can be found at www.cbd.int/convention/.

The Parties, in reaching a *decision on import* under this Protocol or under its *domestic measures* implementing the Protocol, *may* take into account, *consistent with their international obligations*, socio-economic considerations arising from the *impact* of living modified organisms *on the conservation and sustainable use of biological diversity*, especially with regard to the *value of biological diversity* to indigenous and local communities. [CPB 2000, emphasis added]

The negotiations leading to Article 26.1 reflected the contrasting positions on socioeconomic assessments for technology approval processes held by CPB-party countries, nonparties, and other stakeholders in this policy debate. This article has been subject to various interpretations, which have appeared to be drawn mainly from the specific country's position on GM crops (Falck-Zepeda and Zambrano 2011). Note, however, that the CPB restricts the inclusion of socioeconomic considerations as part of the regulatory process to import decisions and implementation of domestic laws and regulations. Moreover, the CPB suggests ("Parties . . . may take into account") the inclusion of socioeconomic considerations, but it does not make its implementation mandatory. As specified in Article 26.1, the application of these considerations should be consistent with the country's international obligations, mainly those to the World Trade Organization. Further, the CPB provides a very narrow scope for the inclusion of socioeconomic considerations, mentioning them only in connection with their impacts on the conservation and sustainable use of biological diversity. Countries have taken dissimilar approaches regarding the inclusion of socioeconomic considerations, from a very strict interpretation of Article 26.1 by Argentina and the European Union to a more relaxed approach by Canada, China, and the United States (Table 1.2).

Given this wide range of interpretations, there is little or no clarity on either which steps are necessary to assess socioeconomic considerations or which methods and tools can best be used to assess these considerations. Equally unclear is how these interpretations can be translated into realistic evaluations. When would be the appropriate time to implement the assessment? Which institution(s) will be responsible for carrying it out? Which entity will be making final decisions regarding the results of the assessments? Table 1.2 summarizes the dissimilar approaches taken by some countries with respect to these points. Aside from these specific questions, any assessments must allow for a limited range of methods and approaches, as data, by

TABLE 1.2   Different approaches to including socioeconomic considerations in biosafety regulatory approval

Issue	Argentina[a]	European Union, Norway, and Switzerland[b]	Canada and United States[c]	Brazil[d]	India[e]	China[f]
Inclusion	Mandatory	Appears to be mandatory	Not required or considered in the biosafety assessment	Required only if a socioeconomic issue is identified in the biosafety assessment	Not required, according to 1989 guideline	Not required
Implementation approach	Sequential	Sequential	—	Sequential	Appears to be sequential	Not defined
Scope	Limited to impact on exports only	Not clear: still being negotiated	None	Not clear	Not clear	Not defined
Implementing agency	Agency in Ministry of Trade	Probably proponent	—	CTNBio and the National Biosafety Council; National Biosafety Council commissioned studies by external consultants	Not clear: the Genetic Engineering Advisory Committee commissioned studies by third parties	Not defined: studies apparently done by third parties
Point evaluation begun	Commercialization	Probably post-release monitoring	Deliberate release or deregulation	Commercialization	Commercialization	Commercialization
Evaluation method	Trade impact assessment	Not clear	—	Not clear	Not clear	Sophisticated impact modeling

Sources: Based on Falck-Zepeda, Wesseler, and Smyth (2010); Pray (2010).

Note: CTNBio = National Technical Commission on Biosafety, Brazil; — = not applicable.

[a]Unofficial policy of approving only products accepted elsewhere has been abandoned. This is an example of a functional system with a delimited socioeconomic assessment approach.

[b]De facto moratorium on genetically modified crops has been in place since 1999. Events of cotton, maize, oilseed, rape, and soybeans are allowed for EU import, whereas only two events, Bt maize MON810 and starch-modified potato, have approval for planting.

[c]Thousands of confined field trials have been approved. At least 16 products have been deregulated. Two major cases have been taken to court.

[d]Rationale for dual bodies was to separate technical assessment from "political" assessments.

[e]Bt cotton only has been approved to date. Socioeconomic considerations do not seem to have ever been a factor in the approvals (Pray 2010).

[f]Socioeconomic considerations seem to have had an impact in terms of supporting approvals (Pray 2010).

definition, would be experimental (drawn from other regions of the world) or just hypothetical.

The inclusion of socioeconomic considerations in the national biosafety regulatory process will also have economic implications for the country. First, the inclusion of such considerations will increase the costs of setting up the regulatory framework and of compliance with it. Second, the inclusion of socioeconomic considerations will probably limit the number of technologies that can successfully be reviewed under the regulatory process and will also curtail the number of technologies released. For these reasons, if a country, after a careful assessment, decides to include socioeconomic considerations in its decisionmaking process, it will be necessary to establish a process that is transparent, feasible, robust, and cost effective (Falck-Zepeda and Zambrano 2011).

This book addresses the demand for guidelines on how to perform assessments of a GM crop technology within the context and the constraints imposed by a regulatory approval process. The monograph has been developed mainly for researchers in developing countries, who often face stringent financial and human resource constraints and at the same time need to quickly respond to policymakers who ultimately determine the future of GM crop technologies in their country.

Ugandan Position on GM Crop Technology

The interest of Uganda in GM crops can be traced back to 2001, when the country ratified the CPB. Since then, the country has continued to participate in discussions about implementation of the protocol. Over the years, several international institutions, projects, and donor agencies have assisted Uganda in developing a functional biosafety regulatory system. These entities include the Global Environmental Facility of the United Nations Environment Programme, the East African Regional Programme and Research Network for Biotechnology, Biosafety and Biotechnology Policy Development, and PBS. The biosafety framework for Uganda was in its final stages by early 2011, with the biotechnology and biosafety policy approved and legislation at an advanced stage. At the same time, the country has been investing substantially in both infrastructure and human capacity for GM product development and in biosafety regulatory processes. To date, the country has managed to implement a number of confined field trials for GM crop materials, including one of GM cotton, using guidelines drawn from the existing Science and Technology Act.

The application for testing GM cotton was preceded by broad stake-holder consultations on the suitability of growing GM cotton in Uganda. This exercise involved many actors in the cotton industry, including the Cotton Development Organisation; the Agricultural Productivity Enhancement Program; NARO; the Uganda National Council for Science and Technology; the Ministry of Agriculture, Animal Industry and Fisheries; the Agricultural Biotechnology Support Project II; ginners; parliamentarians; cotton farmers; exporters; and several development partners. A conference of these parties created the National Taskforce on Cotton Biotech Transfer for Uganda—a key milestone in consensus building. During this process, herbicide tolerance and insect resistance were selected as the traits to be tested. An application was prepared by NARO and Monsanto and presented to the NBC, which approved it in August 2008. Aside from cotton, the NBC has also approved the implementation of confined field trials for banana resistant to black Sigatoka and, more recently, for water-efficient maize and banana resistant to bacterial wilt. The NBC has also approved the notification to import GM sweet potato for contained use.

Scope and Structure

Following the framework described above, this monograph evaluates the impact of GM cotton adoption in Uganda. The farm, industry, and trade analyses are based on the assumption that GM cotton has been approved and that adoption is imminent. In contrast, the analysis of the institutional setting is based on an assessment of the current situation that looks forward to the approval and adoption of GM cotton. Primary information was collected with the use of a household survey (see Appendix 1). This farm-household information was also used to assess the impact on national industries. In addition to using survey questionnaires, informal and semistructured interviews were held with main actors in the cotton value chain as well as with cotton experts (see Appendix 2). The industry and trade analyses draw information mainly from secondary sources and consultations with experts. Given the ex ante nature of the evaluation, information about the market behavior of GM cotton is not available. Instead, we devote a chapter to describing the cotton value chain, and within this framework, we briefly present the cotton market.

There are two important points to emphasize about this study. First, this report is not an adoption study, as such a study would require data on the adoption process and its determinants. The adoption literature is very rich, and the experiences gained through previous adoption studies have helped us with the chapter examining institutional challenges to the potential deployment of the technology in Uganda. Second, information on cotton production in Uganda is widely available, although in some instances, inconsistencies among databases are evident. When these inconsistencies appear, this study gives preference to the information provided directly by the national institutes, such as the Cotton Development Organisation, over other sources.

This chapter has presented the motivation for and research framework of this monograph. It also presents the case study, GM cotton in Uganda, and the scope of the case study's evaluation. The case study examines two GM technologies, Bt cotton and Roundup Ready cotton. Bt cotton is insect-resistant cotton developed to control mainly bollworm attacks. Roundup Ready is an herbicide-tolerant cotton that withstands glyphosate applications targeted to weeds. The next chapter explains the research framework and the data sampling strategy, including the sampling's limitations. Chapter 3 provides a brief historical review of the Ugandan cotton sector and of the seed and product value chain. The chapter also examines the different cotton production systems, giving an overall idea of the main production constraints faced by farmers. Further, it describes the institutional environment that has influenced the approval of GM cotton and confined field trials in Uganda. Chapter 4 analyzes the institutional factors that would influence farmers' decision to adopt GM cotton seed and identifies bottlenecks. The assessment of the potential impact of GM cotton adoption on farmers is presented in Chapter 5, whereas Chapter 6 addresses the impact on the cotton sector. The implications for trade and the possibilities of coexistence of GM and organic cotton are examined in Chapter 7. That chapter also suggests possible changes in the cotton chain that could help minimize potential risks of commingling. Note that commingling can affect the organic cotton market. The last chapter of this monograph outlines the main conclusions and policy recommendations with regard to (1) the potential impact of GM cotton in Uganda and (2) how the methodological framework and tools used in this study can be used in future evaluations.

Chapter 2

Research Framework

José Falck-Zepeda, Daniela Horna, and Patricia Zambrano

An ex ante impact evaluation of any technology poses a number of challenges, and genetically modified (GM) crops are no exception. The most important and evident constraint when evaluating a technology that has not yet been deployed is the impossibility of using primary information to assess the behavior of the technology under farming conditions. In the case of GM crops, the best available data would be the information generated during confined field trials (CFTs), where GM varieties are tested after passing the regulatory approval process. The resulting information from these CFTs, although valuable, has its limitations. Even after a rigorous selection of sites, the whole diversity of agroecological and farming conditions cannot be captured. With limited availability of primary information, the ex ante evaluation has to rely on (1) expert opinions, (2) some assumptions about the technology's future behavior based on its past performance in other locations; and (3) a modest amount of primary information collected from a few sites. This chapter presents a research framework that takes into account these specific limitations that most developing countries face.

Types of Analysis

Institutional Setting

The economic literature on the adoption of GM cotton has underlined the importance of taking into account the role of institutions when analyzing the success of these technologies, particularly in the context of small-scale agriculture (Tripp 2009). The institutional analysis identifies the actors and their roles in the cotton sector's organizational and biosafety framework development. This analysis helps when responding to two main research questions. The first question is how the current institutions and agents affect the commercial approval of GM cotton in Uganda. The second question is how the current institutional settings would enable or hinder farmers' adoption of GM cotton. Social network theory provides the basis for this analysis. Using the Net-Map tool developed by Eva Schiffer (Schiffer 2007), our study maps the main institutional actors, identifies their links, and evaluates their degree of influence. The information on which these maps are built comes mainly from secondary data and semi-structured interviews with different actors and experts involved either with the cotton sector or with the biosafety regulatory process in Uganda.

Farm Analysis

Farmers are the most relevant beneficiaries to be considered in evaluations of technological interventions that seek to improve agricultural production. For this reason, assessment of direct benefits and risks to farmers from a new technology should be done at the farm level. To do such an analysis, the collection of primary information is recommended. Various methods and tools can be implemented, depending on the quality and richness of the available information. Modeling of general and more refined production functions and the use of partial budgets are common methods used for evaluating the impact on farmers of crop biotechnologies (Shankar and Thirtle 2005).

In our study, the farm evaluation is done using partial budgets and stochastic analysis. Given that GM cotton has not yet been planted in Uganda, the study is based on assumptions about relevant factors, such as GM-crop prices and adoption rates. These factors are crucial in evaluating future economic benefits. The instrument used for the farm analysis is a survey that enables (1) the calculation of partial budgets for representative growers and (2) the simulation of partial budgets for various scenarios, including the use of insect-resistant and herbicide-tolerant varieties. The use of different scenarios that reflect the variability of GM crop prices and other factors gives an idea of the scope of the GM crop's impact. The simulations run include an organic production system with premium price, a hypothetical case in which insect-resistant (Bt) seed is used in an organic system, the adoption of Bt seed, and the adoption of HT seed.

Industry and Sector Analysis

It is important to evaluate which sectors of the economy will benefit and which sectors will lose from the introduction of a new crop technology. Production technologies that are widely adopted can have an impact on demand and supply of the targeted crop. This impact can have an effect on commodity prices and on the distribution of benefits between producers and consumers. To estimate the effect on the different sectors of the economy, our study uses the economic surplus method, a widely used partial equilibrium approach (Alston, Norton, and Pardey 1995). For the implementation of the approach, Uganda is classified as a small open economy, because the country's cotton trade has no influence on the international price of cotton lint. Thus, the new production technology will only influence supply—that is, it will only change producers' surplus.

To predict the effect of GM cotton introduction, it is necessary to account for all variables affected when moving the technology from the research and development phase to the hands of farmers. These variables are often expressed in terms of prices and costs, but such factors as adoption levels, durability of resistance, and time lags caused by biosafety processes and adoption lags also need to be considered. In an ex ante assessment, it is necessary to make assumptions regarding the size and value of each of these variables, as the actual numbers are usually unknown. Multiple scenarios are developed to assess the robustness of these assumptions. To consider a range of productivity impacts that might be observed on farms, the study moves from the basic economic surplus approach with one set of values per variable to a distribution of values. This approach enables us to address variability and risk, which characterize the adoption of any new technology (Falck-Zepeda, Traxler, and Nelson 2000; Falck-Zepeda 2006).

Trade Analysis

The evaluation of the impact on international trade is particularly important in the case of GM technologies, given the strong regulations of some non-adopting commercial partners, particularly European countries. Different approaches, from simple bilateral flow analysis to complex trade models, can be used to assess the potential effects of GM crop adoption in the context of trade regulations. Because the main product of cotton is lint, GM cotton is not subject to stringent import regulations in Europe or Asia as are food crops. In fact, GM cotton is mixed with conventional cotton in the global cotton lint market. Instead of regulation-related issues, published studies have focused on the effects of nonadoption when competitors adopt and the combination of adoption with other policy considerations. For instance, Elbehri and MacDonald (2004) use a multiregional computable general equilibrium model to assess and contrast the welfare effects of non- and partial adoption of Bt cotton in West and central African countries. Anderson, Valenzuela, and Jackson (2008) combine GM cotton adoption with trade liberalization scenarios to quantify the gains to African countries if trade liberalization policies were implemented and followed by GM cotton adoption.

The goal of this trade analysis is to provide a rapid assessment of the possibilities for and potential consequences of the coexistence of organic and GM cotton in Uganda. Because Uganda is a small cotton-producing country and trade models use regionally aggregated databases, our study does not provide an economic simulation of GM cotton adoption in Uganda. Instead, it combines some results from the literature with data based on the assessment of organic cotton's coexistence and segregation. Whether and how organic and GM cotton can coexist is the most important trade-related issue for GM cotton introduction in Uganda. As described later, Uganda has a small but growing organic cotton sector. Drawing on a review of secondary information and the literature available, our analysis provides a rapid overview of constraints on and opportunities for GM cotton adoption, and offers insight into how both organic and GM cotton growers can produce independently for the benefit of all.

Data Collection Instrument

A survey was used to collect information on cotton production and current practices (see Appendix 1). In addition to questions about input use and production, the survey asked growers about yield distributions to gauge farmers' perception of the extent of yield losses caused by the constraints of bollworm and weeds. The triangular distribution (minimum, maximum, and mode) is the simplest one to elicit from farmers, approximates the normal distribution, and is especially useful in cases where no sample data are available (Hardaker et al. 2004). Farmers were asked for their highest, most common, and lowest yields, with and without the constraints. This question was asked first for bollworm and then for weeds, which are the constraints targeted by Bt cotton and herbicide-tolerant cotton.

Lira and Kasese were the only two districts selected by the National Biosafety Committee to implement the CFTs, despite the highly heterogeneous conditions of cotton cultivation. The districts were selected after careful deliberation about the economic relevance of cotton production and safety considerations. Kasese is an important cotton-producing and processing district, and Lira is the main district producing organic cotton. We speculate that a limited budget was also an underlying criterion in the selection process.[1] Given that the CFTs were done in the context of a biosafety regulatory

---

1   Limited resources to implement socioeconomic evaluations is probably a constraint for any local team trying to evaluate the ex ante impact of GM technologies and contribute to decisionmaking in a developing country. We decided to mimic these constraints as much as possible in our work.

process, we decided to implement the household survey in the same districts. Although the selection of these districts was for important economic and bio-safety reasons, we understand that it also affects the statistical significance of the sample.

First, local extension agents used their field experience to identify villages as "cotton producing." Next, we randomly selected three of these villages in Lira and seven in Kasese. The relative shares of villages selected in Lira and Kasese reflected the proportion of cotton produced in those districts as well as the need to have good representation of organic producers. A total of 150 household heads were interviewed, 30 in Lira and 120 in Kasese. Households were randomly selected from a list of producers provided by ginneries operating in each district. The number of households selected from a village followed a qualitative evaluation by local extension agents of cotton produced in each village. This procedure was followed because there are no official village cotton-production data by district level. When it was not possible to interview the selected household, a new name was randomly selected from the available list of producers.

The questions in the household survey were addressed to the household head for the 2007 growing season, and some additional information was collected for 2006 to correct potential mistakes in data entry and missing information in data collection. In some cases, especially in villages in Lira, selected producers cultivated more than one plot. The production information was analyzed per plot rather than per producer for the 2007 growing season only. Plots with incomplete information were not considered in the analysis. Thus, the total number of observations in our analysis ended up being 151 plots; of these, 35 were plots from producers in Lira (12 plots from organic producers), and 121 were plots from producers in Kasese.[2]

Several constraints during field implementation affected our sample and induced potential biases:

- Not all producers in Lira were organic producers despite being self-identified as such; as a result, the organic producers sample ended up being very small.

- The list of producers provided by ginneries, despite being the current and updated one, was incomplete.

---

2 Five observations were dropped because of incomplete information. These observations corresponded to five plots from Lira.

- Incomplete information after data collection forced us to drop observations, affecting attrition and the representation of producers at the village level. The observations dropped, however, were only in Lira.

Considerations for Data Collection

Representativeness

The relevance of the socioeconomic assessments relies on the quality of their results. Quality is measured not only by the robustness of the methods used to reach the results but also by the degree to which these results help derive policy recommendations at the national level. To be able to make inferences at the country level, the sample selected has to be representative of the national conditions.

To examine representativeness, it is necessary to assess whether the production from both selected sites represents the overall cotton production in Uganda and to review the current spatial distribution of cotton production in the country. Table 2.1 shows the cotton area and yield estimates in Uganda by district. Lira and Kasese together represent approximately 22 percent of the total cotton area (Lira is roughly 14 percent and Kasese 8 percent). The information in this table is fairly consistent with other figures and data presented elsewhere in this book. Given the relatively large share of both areas and the sampling process (in which villages and households in a district were selected in proportion to the percentage of national cotton production produced by that district), data taken from these districts are likely to provide a relatively good estimate of national seed cotton yield and pesticide cost variations. This is particularly true when national statistics are expressed as probability distributions in simulations to assess the uncertainty of model parameters, as is the case in this book.

Representativeness can also be assessed by comparing the parameters defining our sample with the parameters of national statistics or alternative recognized data sources. FAO agricultural statistics and the Uganda National Household Survey (UNHS) (UBOS 2007) are probably the most complete panel datasets available. These datasets are very useful for country-level analysis but are of limited use for more disaggregated analysis; this is particularly true of the FAO data. The UNHS contains more disaggregated data than the FAO statistics, but exactly how much disaggregated data is available depends on the crop under evaluation: some crops have more detail available than do

TABLE 2.1 Ugandan cotton area and yield by district, 2000

District	Cotton area (hectares)	Cotton yield (kilograms/hectare)	Share of cotton area (percent)
Pallisa	42,871	262	17
Lira	36,219	273	14
Nebbi	33,302	272	13
Apac	27,325	266	11
Kasese	19,761	272	8
Tororo	9,515	270	4
Kamuli	9,466	268	4
Masindi	7,378	275	3
Busia	7,270	270	3
Iganga	6,388	270	3
Sironko	6,256	281	3
Pader	5,534	271	2
Kitgum	5,129	270	2
Kaliro	4,233	275	2
Bushenyi	4,044	274	2
Butaleja	3,865	270	2
Amuria	3,130	265	1
Other districts	18,311	265	7
Total	249,999	—	101[a]

Source: IFPRI and HarvestChoice (2012).

Notes: Numbers are rounded estimates. Districts with fewer than 3,000 hectares were aggregated into the "Other districts" category in the table. The yield shown for the "Other districts" is the average for all districts included in this row. — = not applicable.

[a]Percentages add to more than 100 because of rounding.

others. You and Chamberlin (2004) used UNHS data to develop spatial estimates of cotton production areas based on altitude, growing period, population density, and market accessibility estimates. All this information was later translated into two-dimensional maps. The authors found that the relatively low number of cotton-producing households in the UNHS limited not only the ability to map cotton production areas in greater detail but also the ability to describe heterogeneity within the cotton production zone and farmer heterogeneity.

You and Chamberlin (2004) drew conclusions that suggest that both Lira and Kasese were important cotton production areas. Most important, they projected that future areas for intensification and expansion would be in or around Lira or areas with the same cotton production domain. Note that a production domain is characterized using agroecological, demographic, accessibility, and market information. Even though our survey captured a relatively low number of responses from Lira, the large number of responses from Kasese, a district that represents a seemingly different production domain, compensates for this limitation.

Farmer Heterogeneity

The importance of producer heterogeneity for the estimation of adoption impacts has been documented by Feder, Just, and Zilberman (1985) and Sunding and Zilberman (2001). Farmer or household heterogeneity is a direct result of the existing conditions under which the producing unit operates. These conditions include socioeconomics, market access, and biophysical conditions and are related to the ability to gain value from technology adoption. In practice, spatial differences in the impact of Bt cotton may occur because of agroecological conditions or the production practices of farmers (Kathage and Qaim 2012).

To examine how good data from our survey would be in providing a gross estimate of broader national heterogeneity in Uganda, it is important to examine the internal consistency and variation of the data in the sample used to implement the survey. We examined the percentage distributions of both yield and pesticide costs to control lepidopteran insects in our Uganda survey (Figure 2.1 and Table 2.2).

The data are fairly consistent, as Kasese produces on average almost double the amount of seed cotton produced by Lira. Lira also incurs significantly lower costs for applying pesticide that targets bollworm and many kinds of lepidoptera that GM cotton is supposed to resist. Further, Figure 2.1 shows a fairly uniform yield distribution in our sample across different quartiles, implying that a probability distribution will be able to grossly approximate potential outcomes in repeated iterations in a simulation. The same conclusion does not apply to the pesticide costs, because few farmers reported any pesticide use to control lepidopteran insects in Lira (Table 2.2). In fact, only 7 of the 34 farmers surveyed in Lira reported using insecticide. Because we are using continuous probability distributions in our simulations, this characteristic of the data might bias results (unnecessarily sampling from the tail). Yet it may not be a significant limitation in our study, as we aggregated producers from both Lira and Kasese to derive probability distributions that are broad enough to encompass a wide variation of outcomes.

FIGURE 2.1   Percentage distribution, by quartiles, of seed cotton yield
in Lira and Kasese, Uganda, 2007

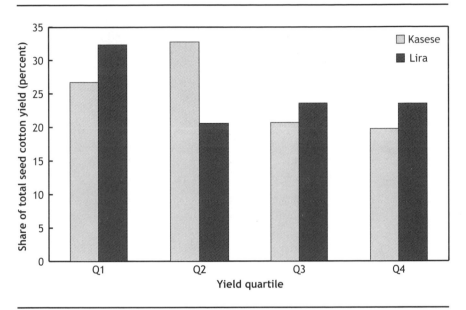

Source: Authors' estimations based on field survey.

TABLE 2.2   Summary of cotton yield and pesticide costs to control lepidopteran insects
in cotton, Kasese and Lira, Uganda, 2007

	Yield (kilograms/hectare)		Pesticide costs to control lepidopteran insects (US$/hectare)	
Indicator	Kasese (N = 116)	Lira (N = 34)	Kasese (N = 116)	Lira (N = 34)
Average	1,038	696	8.8	2.4
Minimum	49	86	0.0	0.0
Maximum	4,098	2,059	39.0	32.0
Standard deviation	746	544	7.2	6.8
Median	988	515	7.0	n.a.
Mode	988	1,235	14.0	n.a.
Skewness	1.3	1.3	1.6	3.5
Kurtosis	2.8	0.9	3.5	12.5

Source: Authors' survey data.
Note: n.a. = not available.

To investigate a finer disaggregation of outcomes stemming from spatial heterogeneity, farmer heterogeneity, or both—for example, differences in farm size—or to consider multiple agroecological zones or geographical demarcations, we would need a sufficient number of observations to derive a probability distribution for each subgroup. For instance, if there are four main agroecological zones where cotton is produced, we would need a representative number of observations for each zone. The relatively small number of respondents in our survey did not allow this option. However, this limitation may not be as binding as it might appear, particularly because most of the cotton production area is located in the humid and subhumid agroecological zones where Kasese and Lira are located. But in essence, our results can only be considered as gross estimations of spatial heterogeneity, farmer heterogeneity, or both that require more detailed information about probability distributions to assess heterogeneity more precisely. This is a major limitation of our study.

Chapter 3

The Cotton Sector in Uganda

Daniela Horna, Patricia Zambrano, and Theresa Sengooba

Ugandan farmers have planted cotton for more than a century. Despite this long tradition of cotton cultivation, cotton productivity in Uganda is very low compared to international and regional averages. During 2003–07, the average seed cotton yield was 428 kilograms per hectare, among the lowest in Africa and in the world (FAO 2010). This chapter presents the main historical developments of cotton production in Uganda; it is based on Cotton Development Organisation (CDO) reports (CDO 2006) and a literature review. The value chain for seed cotton, cottonseed, and lint is also discussed. This chapter also describes the production systems and identifies the main constraints on cotton production. The last part of the chapter evaluates the institutional setting and the biosafety regulatory process. The objective of this chapter is to provide a comprehensive picture of the past and present situation of the Ugandan cotton sector. This information feeds into the evaluation of the potential impact of genetically modified (GM) cotton adoption and helps identify bottlenecks and research needs.

Cotton in Uganda

Historical Background

Cotton cultivation in Uganda dates to 1903. The history of the crop and its performance is well documented in the literature (You and Chamberlin 2004; CDO 2006; Baffes 2009; Tschirley, Poulton, and Labaste 2009). Cotton cultivation was originally concentrated in the central part of the country, but over time such food crops as coffee and bananas displaced this crop. Currently, cotton is cultivated mostly in Eastern, Northern, and Western Regions, where other cash crops have only limited potential. According to Baffes (2009), these areas now account for roughly 60 percent of total production.

From the 1950s until 1970, cotton, along with coffee, became the most important source of revenue for the government. By 1960/61, Uganda was producing 218,000 tons of seed cotton, reaching its historic peak in 1970

(Figure 3.1). After independence in 1962, the government set up the Lint Marketing Board to have exclusive control of all lint and cottonseed. This decision, along with highly favorable international prices, generated a considerable increase in production, reaching in 1969/1970 the highest output ever produced in Uganda, 470,000 bales of lint. This outstanding production was mainly due to area expansion rather than increases in productivity.

The period of civil unrest from 1971 to 1979 ruined the country and affected all its productive sectors, including the cotton industry. Crops were abandoned, and as a consequence, both the area under production and productivity dropped dramatically. Seed cotton yield dropped from 320 kilograms (3,240 hectograms) per hectare in 1974 to less than 50 kilograms (428 hectograms) per hectare in 1980. The dramatic decline in productivity was also the consequence of a failure to maintain and multiply existing cotton seed varieties and of poorly maintained ginning operations, among other reasons (Baffes 2009).

In 1986, a new government came into power and embraced macroeconomic reforms to promote economic development. Government institutions were sold, privatized, or restructured. Restructuring of the cotton sector was carried out from 1994 to 2000 with the support of the World Bank's International Development Association and the International Fund for Agricultural Development. The Cotton Development Act established

FIGURE 3.1   Cotton production in Uganda

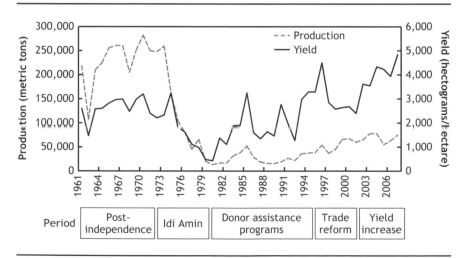

Sources: FAO (2010); Baffes (2009).
Note: 10 hectograms = 1 kilogram.

cotton zones in 2003 and restricted the movement of cotton into or out of those zones. Producers were then forced to bring their output to the ginneries located within their cotton zones (CDO 2004). In 2006, cotton producers were given the right to sell their cotton to the highest bidder, making it harder for ginneries to control the supply of cotton and diminishing ginneries' interest in providing extension services to farmers. Ginnery operators argue that they no longer have incentives to provide extension and agricultural inputs to farmers, because farmers can sell their cotton to a different ginnery.

Cotton producers are still dependent on inputs supplied by ginneries: mainly seed and occasionally fertilizers and insecticides. Uganda's National Agricultural Advisory Services provide limited extension services, and technology development is in the hands of the National Agricultural Research Organisation (NARO). CDO continues to play a crucial role in the sector as the unique seed provider and a key player in the commercialization of cotton.

Despite the efforts made by the government to revitalize the sector, cotton production has grown very little over the past decade. It is apparent that other economically important crops, such as plantains, beans, and maize, now occupy areas that were devoted to cotton (Figure 3.2). In recent years, rice production in Eastern Region's districts has become a more profitable economic activity for farmers (UEPB 2007). In addition to these competing crops, the cotton sector faces new challenges. To begin with, Uganda no longer commands a premium price for its better cotton fiber quality. Although there is only one variety of cotton under cultivation, the lack of cotton lint uniformity caused the loss of premium price in 2002. Although environmental and soil

FIGURE 3.2   Production of main crops since 1970

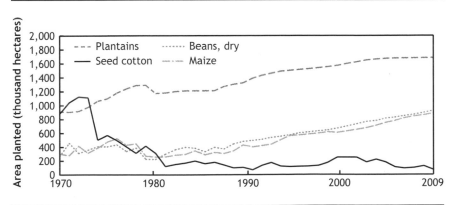

Source: FAO (2010).

properties favor the development of the crop across the country, climatic conditions have contributed to the slow recovery of the sector. Drought periods followed by excessive rain, low temperatures, and cloudy weather are considered primary causes of low cotton yields. Recent increases in productivity have been the result of NARO selecting cotton varieties, CDO distributing credit and inputs to farmers, and extension work improving agronomic practices.

Current Cotton Production

According to Baffes (2009), there are approximately 250,000 low-income cotton households in Uganda. An earlier estimate by Gordon and Goodland (2000) puts this number between 300,000 and 400,000. Both studies recognize that this crop can improve farmers' welfare. At the aggregate national level, however, cotton accounts for just 2.2 percent of all crop production and thus plays a relatively small part in rural livelihoods (You and Chamberlin 2004). With respect to trade, cotton ranks third among agricultural commodities exported, although it accounts for only 2–5 percent of Uganda's total exports (Serunjogi et al. 2001).

Cotton is cultivated in more than 30 districts across Uganda because of the favorable agroclimatic conditions for the development of the crop (Figure 3.3) (You and Chamberlin 2004). The important producing areas are located in Northeastern, Southeastern, Northern, and Western Regions (Table 3.1). Figure 3.4 presents the production of seed cotton in 2006 by subregion and district. Note that Kasese District in Western Region was the most important seed cotton producer in 2006.

Cotton Value Chain

As in other West and East African countries, the cotton sector in Uganda is characterized by vertical integration and strong public sector involvement (Figure 3.5). Poulton and Tschirley (2009) have classified the African cotton sectors into five types based on the structure of the market for seed cotton purchases and the regulatory framework in which purchasing firms operate. In Uganda, the sector is considered a hybrid type, as the market is regulated but no national monopoly exists, which makes it possible for firms to buy seed cotton from different geographical areas.

Seed Value Chain

Often the successful adoption of an agricultural technology in developing countries depends on the ability of the public sector, private sector, or both to

FIGURE 3.3   Cotton regions in Uganda

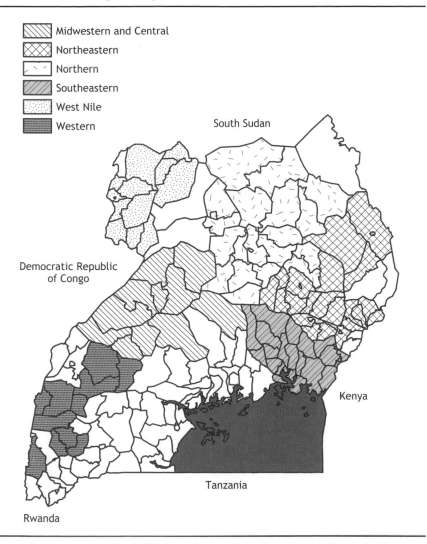

Source: Mapped by Ulrike Wood-Sichra of IFPRI, using designations of cotton-producing zones and districts from CDO (2006).

provide adequate amounts of good quality seed in a timely manner. In Uganda, the performance of cotton production in the field depends on the availability of good quality seed. The use of poor quality seed has evident repercussions throughout the cotton chain. Seed production and manipulation in Uganda, although regulated, do not yield the best quality, and often the volumes produced do not cover farmers' demands.

TABLE 3.1   Cotton-producing regions in Uganda

Region	Districts	Number of households involved in crop husbandry
Southeastern	Bugiri, Busia, Butaleja, Iganga, Jinja, Kaliro, Kamuli, Kayunga, Mayuge, Namutumba, Tororo	550,637
Northeastern	Budaka, Bukedea, Bukwa, Kaberamaido, Kapchorwa, Katakwi, Kumi, Manafwa, Mbale, Moroto, Pallisa, Sironko, Soroti	535,413
Northern	Amolatar, Amuria, Apac, Dokolo, Gulu, Kitgum, Kotido, Lira, Oyam, Pader	532,348
West Nile	Adjumani, Arua, Moyo, Nebbi, Nyadri, Yumbe	323,584
Western	Bushenyi, Kamwenge, Kanungu, Kasese, Kyenjojo	369,008
Midwestern and Central	Bulisa, Hoima, Kabale, Kiboga, Luweero, Masindi, Nakasongola, Sembabule	379,390
Total		2,690,380

Source: CDO (2006).

FIGURE 3.4   Seed cotton production by district, 2006

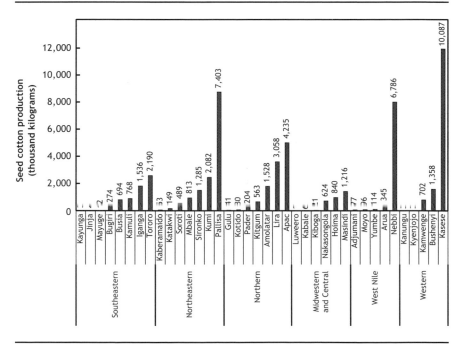

Source: ACE (2006).

FIGURE 3.5   Cotton value chain in Uganda

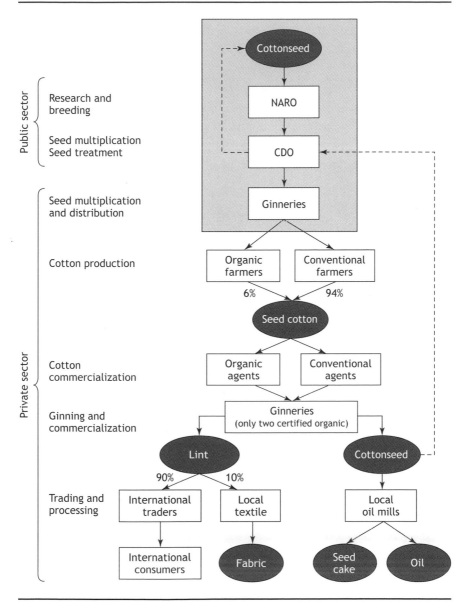

Sources: Based on information provided by CDO, NARO, and ginneries representatives.

Note: CDO = Cotton Development Organisation; NARO = National Agricultural Research Organisation.

The main actors in the cotton seed chain in Uganda are cotton producers, NARO, CDO, and some private ginneries. NARO coordinates and oversees all aspects of agricultural research in Uganda and is in charge of cotton breeding research and cotton technology development.[1] The need for improved varieties and certified seed is probably the most important constraint encountered in cotton production in Uganda (Serunjogi et al. 2001). NARO and the institutions that preceded it (before and after independence) were relatively active in selecting and releasing improved cotton varieties. However, the multiplication and seed distribution process needs more attention. Note that the sector has been liberalized, but cotton seed production is nevertheless entirely channeled through the ginneries, NARO, and CDO.

CDO is likely the most important agent in the cotton chain.[2] It regulates, coordinates, and promotes all aspects of the cotton subsector in Uganda. CDO also monitors cotton production and marketing, provides policy advice regarding the crop, and makes a rough estimation of the seed volume needed for each season's production campaign (CDO 2006). Basic cotton seed is developed by NARO but multiplied by CDO and distributed mainly to the ginneries for commercial multiplication and further distribution to farmers. Therefore, new cotton seed is produced every year, primarily in Kasese, the most important seed-producing district.[3] Often the new seed does not cover the demand for seed, and some seed is recycled in the system from previous campaigns. The recycled seed is also under the control of CDO. At harvest, after delinting the seed cotton, ginneries are required to save the best quality cotton seed for CDO.[4] The ginneries can use the rest of the cotton seed or sell it to oil and milling companies. CDO is in turn responsible for delinting, grading, and dressing all seed.

---

1 NARO was established by an act of parliament on November 21, 2005. See www.naro.go.ug/About%20NARO/aboutnaro.htm.

2 CDO was established by an act of parliament in 1994 (http://cdouga.org/). It has the responsibility to monitor the production, processing, and marketing of cotton so as to enhance the quality of lint exported and locally sold, to promote the distribution of high-quality cottonseed, and to facilitate generally the development of the cotton industry. CDO mobilizes and encourages farmers to form farmer groups for easier delivery of services and for access to credit and collective bargaining at marketing time. It has assisted in the formation and development of a private sector body: the Uganda Ginners and Cotton Exporters Association. The ginners support cotton production and productivity, and their strategy focuses mainly on the registration of cotton farmers in their areas, thus creating a reliable database to ensure that planting seed is supplied only to genuine cotton farmers and seed wastage is curbed (Muwanga-Zake 2009).

3 To reduce the cost of transporting planting seed from Kasese to other regions, CDO, in 2002, reactivated segregated areas for seed multiplication in each region.

4 Seed cotton is the product harvest, which includes lint still attached to the seed. Cottonseed is the delinted seed cotton and cotton seed is the seed used for planting.

The availability of cottonseed is a very limiting constraint on improving cotton productivity in Uganda. According to the Uganda Export Promotion Board (UEPB 2007), cotton exports in 2006 fell 29 percent compared to the previous year. The low performance in 2006 was the result of several factors, including late planting. However, the main reason for the low performance was the use of ungraded fuzzy (that is, not delinted) seed, which led to high seed wastage and higher costs of seed provision (CDO 2006). In general, ginning equipment is outdated, which can adversely affect the final quality of the seed. CDO has intervened to support efforts in delinting and seed grading, but the problems persist. Investment in postharvest equipment could have a significant impact on ginning turnout and thus on the final profitability of cotton.

Ugandan production is characterized by the availability of only one seed variety, which is used for both conventional and organic planting. This is a long-staple variety called the Bukalasa Pedigree Albar, which is distributed by CDO free of charge. The one-seed policy has been promoted as a way to guarantee quality homogenization, because it is believed that a single variety ensures uniformity in the production of lint and yarn. However, the dependence on a single variety increases the vulnerability to pests and diseases and represents a potential risk. Note that the current Bukalasa Pedigree Albar variety is susceptible to bacterial blight, fungal wilts, Lygus sap bug, pink bollworm, spiny bollworm, American bollworm, aphids, whiteflies, and cotton-staining buds.

Product Value Chain

Farmers, intermediate agents, ginners, exporters, and CDO are the main actors in the cotton product value chain. It has been common for ginners to provide farmers with fertilizers and insecticides, in addition to seeds. Farmers pay back at harvest time either with cotton or cash. The amount of inputs used is still limited, however, probably because farmers plant cotton not as a commercial enterprise but as a secure cash crop they can count on—cotton has a set price and will eventually bring in income after harvesting. This is particularly true in the northern and eastern part of the country, where smallholder farmers have limited cash production alternatives to cotton, either as a single crop or as part of a rotation (Baffes 2009).

At harvest time, farmers can take their production to the ginneries, but often the small volume produced does not justify paying for transport. Most often, intermediary agents, who can work independently or for specific ginneries, gather cotton production from several producers and take it to the ginneries. There were originally restrictions on the production areas that a ginnery could cover, and ginning companies were allowed to buy and process only cotton produced in their neighboring areas. As farmers can currently sell their production to the agent who offers them the highest price, competition among agents has increased. Appendix 3 presents the activity of ginneries by region in 2007.

More than 50 ginneries, owned by approximately 27 different private companies, are distributed across the country (see Appendix 3 for more information). Ginneries compete for access to cotton areas because, given the relatively low cotton production, most ginneries operate below their potential capacity and are active only during the harvest period. The ginning machines are of poor quality, and consequently their turnout is low. Dunavant (U) is the largest company, with six operating ginneries, and accounts for 16.6 percent of cotton lint production. Dunavant is also the only company that has certified organic ginneries. The other organic ginneries, Copcot and Lango Cooperative Union, segregate areas for organic production when the organic cotton supply is sufficiently large to warrant separate areas.

The local industry consumes approximately 7–10 percent of the lint produced; the rest is exported to international markets. In 2007, cotton exports were valued at 36 million US dollars; Switzerland, the United Arab Emirates, Singapore, and the United Kingdom accounted for 82 percent of revenues (FAO 2010). All organic cotton is exported to Europe, mainly to Switzerland.

CDO, in collaboration with the ginners association, is in charge of determining and fixing the farmgate price of seed cotton, which fluctuates during the marketing season (Baffes, Tschirley, and Gergely 2009; FAO 2010). CDO has not established an enforcement system for the farmgate price, however. For the past eight years, this price has been set at about 50–70 percent of international cotton lint prices, as measured by the Cotlook A index (Tschirley, Poulton, and Labaste 2009). To this estimated farmgate price, ginneries have to discount the ginning costs and small charges for transport, as applicable. Because the Cotlook A index is taken into account to fix the farmgate price, there is a direct relationship between farmer profits and world price variability. As world prices are greatly dependent on production variability among the main cotton producers, the adoption of GM cotton and the associated increases in volume of production by large producers, such as China, India, and the United States, have had important implications for the Ugandan cotton sector. A more in-depth discussion of these effects is presented in Chapter 6.

Cotton Production Systems

Traditional System

Cotton production in Uganda is mainly in the hands of smallholders and is characterized by low input use and low profitability. Small-scale producers are caught in a food insecurity trap and work for immediate cash rather than trying to improve their cotton fields (Poulton, Labaste, and Boughton 2009). Not only are inputs expensive and unaffordable, but also their availability is limited. Access to such production inputs as fertilizers, or even good quality seed, is difficult to predict. Usually, the cotton season starts only when seed is made available to farmers by the ginneries. Given the vertical integration and high articulation among actors in the value chain, delays in the delivery of seed and other inputs commonly occur because of bureaucratic sluggishness or bottlenecks.

According to Poulton, Labaste, and Boughton (2009), in addition to small-scale low-input farmers, there are some large-scale cotton farmers in Uganda who achieve yields of more than 2,000 kilograms per hectare. These large-scale farmers have the highest productivity among comparably sized farms in cotton-producing countries in Africa south of the Sahara. Despite these highly productive farmers and the fact that in the past five years Ugandan seed cotton yields have registered their highest levels since 1960, the country's average cotton productivity is below international and regional averages. Poulton, Labaste, and Boughton (2009) explained that the significant difference between large- and small-scale producers can be explained by two main factors: (1) access to inputs and (2) ownership of assets.

Cotton production in Uganda has shown great variability over the years (see Figure 3.1). In the 2006/07 campaign, Uganda produced 75,000 metric tons of seed cotton, recording an average yield of just 483 kilograms per hectare (FAO 2010).[5] This production resulted in a total of 24,790 tons of lint (CDO 2006). Even though the area cultivated increased from 100,000 hectares in 2005/06 to 150,000 hectares in 2006/07, yields declined in the same period. In the 2007/08 campaign, total cotton production sharply declined. Total lint produced was only 12,303 tons—merely 65 percent of the amount produced the previous season.

Several factors cause variability in production. External events, such as drought or excessive rain during growing periods, contribute significantly to variability. Because cotton is a rainfed production system, climatic events,

---

5 All tones are metric tons in the chapter.

especially variability in precipitation, can severely affect cotton yield. In the 2006/07 season, for instance, rains were not timely, some areas were hit by hailstorms during crucial production stages, and other regions lacked enough moisture in the soil for cotton boll formation. Institutional factors are also crucial determinants of this high output variability. CDO acknowledges that the poor availability of high-quality inputs, including seed, extension, and credit, are reasons for low performance. The situation became particularly acute in the 2007/08 season, when the ginneries dropped their production support program (CDO 2008).

Organic System

Uganda plays an important role in regional organic production, as it appears to be the country with the largest area planted with organic cotton in Africa. Together with Tanzania, Uganda was among the first organic cotton producers on the African continent (Ogwang, Sekamatte, and Tindyebwa 2005; Moseley and Gray 2008). The initial promoters of organic production saw an opportunity to establish their cotton operations in the Lango districts (eight districts in the Northern Region; see Figure 3.3). Because small-scale farmers in these districts had traditionally used little or no insecticide, it was easier for them to run an organic operation without making any major adjustment to their practices.[6] In fact, as many farmers cannot afford to invest in insecticides and other chemicals in the first place, an organic cotton system is a continuation of usual practices. Under these circumstances, farmers are not necessarily thinking strategically about the potential niche market for organic crops or about environmentally harmonious production practices. Instead, they are opting to be organically certified in an attempt to increase their low profit margins. This can have an effect on the commitment to organic production and continuity with organic practices. See Appendix 4 for a brief history of organic cotton cultivation in Uganda.

As in any other organic agricultural system, the Lango system is characterized by the use of biological control and agronomic practices to control pests and diseases, limited use of productivity-enhancing technologies, and certification of the whole farming system and individual ginneries. In Lango, cotton is produced in rotation with sesame, an oilseed crop that commands a much higher productivity and market price compared to cotton. Farmers do not

---

6 In 1994, the organization Export Promotion of Organic Products from Africa, a Swedish development initiative, began the Lango Organic Project in Lira and Apac Districts.

have many alternatives to cotton in this rotation system. Table 3.2 shows the intense expansion of organic cotton production since 2000.

From 2000 to 2006, organic cotton lint production made up, on average, 2.5 percent of total annual cotton lint production (CDO 2008). This share of national production increased to more than 9 percent in the 2006/07 season. It appears that the growing interest of farmers in a crop system that requires lower use of often-unaffordable inputs and that could possibly command a premium price has motivated this fast increase in area planted with organic cotton. In 2007/08, the production share of organic cotton was even higher, more than 20 percent (Table 3.2), showing a still-growing interest in the organic system. However, these numbers do not reflect the low productivity of cotton during the season. There are no statistics available for areas planted with organic cotton over the seasons, so it is difficult to quantify the sector's performance over time. CDO representatives explained that although organic cotton areas have expanded at a rapid pace, the overall national cotton yield in the 2007/08 season was particularly low. Thus, the contribution of organic lint to total lint production was particularly high, despite the unfavorable production results.

Institutional Setting and the GM Regulatory System

The success of a technology can be limited or enhanced by the institutions that support it. Before GM crops are approved for commercialization,

TABLE 3.2   Organic cotton production, 2000–2006

Growing season	Organic seed cotton production (kilograms)	National seed cotton production (kilograms)	Organic lint production (bales)	National lint production (bales)	Organic lint's share of national production (percent)
2000/2001	1,642,458	54,996,904	3,066	102,200	3.0
2001/2002	1,734,187	63,898,025	3,406	126,148	2.7
2002/2003	1,203,753	57,563,429	2,407	114,619	2.1
2003/2004	2,030,465	84,344,870	3,626	160,000	2.3
2004/2005	2,979,969	130,854,714	5,321	254,000	2.1
2005/2006	1,499,030	51,847,138	2,677	102,000	2.6
2006/2007	7,377,333	68,681,469	13,174	134,000	9.8
2007/2008			13,766	66,500	20.7

Source: CDO (2008).
Note: 1 bale = 185 kilograms.

products need to go through the national regulatory biosafety system. Ideally, this biosafety system is in place and effective when the technology is released to ensure both consumer protection and timely access to new technologies. In reality, biosafety systems in many developing economies are likely to be under development and involve multiple actors entangled in complicated and lengthy processes when GM crops are released.

As in other countries at similar stages of development, the biosafety regulatory framework in Uganda is still under development. Two critical requirements for moving GM cotton approval forward are (1) the approval of the confined field trials (CFTs) and (2) the approval of the Biosafety Bill. CFTs are small-scale field experiments where the performance of GM plants is evaluated under stringent conditions. The Biosafety Bill is a national legal framework that sets the rules for the handling and use of GM crops and also complies with the requirements of the Cartegena Protocol on Biosafety.

Approval of CFTs

The implementation of CFTs gives technology developers and decision-makers the opportunity to evaluate GM crop performance. These trials are also the opportunity to collect data required for the purposes of safety assessment, variety testing, registration, and seed certification. In a country with limited experience with CFTs, the approval process can be a very lengthy one, as it will require learning and building expertise. Uganda's only previous experiences with CFTs were the trials for fungus-resistant GM bananas implemented in 2007.

In August 2007, the National Biosafety Committee of Uganda (NBC) issued a permit for the implementation of CFTs for insect-resistant cotton varieties and herbicide-tolerant cotton varieties. The CFTs took place in Mubuku, Kasese, and the National Semi-Arid Resources Research Institute, Serere. Before these field trials were implemented, NBC required applicants to answer a round of questions that was followed by preparation of trial sites. The inquiry, preparation, and approval process for the CFT took almost two years, which pushed the starting date to June 2009. The complexity of the process is shown by the direct or indirect involvement of at least 21 actors and agencies in the processes that led to the approval of CFTs (Table 3.3).[7]

---

7 Although the study initially identified at least 21 actors with possible influence on the process (Table 3.3), the application of Net-Map techniques enabled the researchers to illustrate the approval process with just 9 actors, to evaluate each actor's degree of influence, and to identify bottlenecks in the process.

TABLE 3.3   Actors that can influence the establishment of genetically modified (GM) cotton-confined field trials (CFTs)

Institution	Type	Role	Influence
Cotton Research Institute	Public	Design and implement field trials	Direct
Uganda National Council for Science and Technology (UNCST) and UNCST Executive Secretary	Public	Ensure biosafety in testing and development of GM crops and provide support for training, capacity building, regulatory strategies, and policy development	Direct
National Biosafety Committee (NBC)	Public/Private	Provide regulatory oversight and enforcement on behalf of citizens in all matters concerning biosafety in the research, development, and utilization of GM organisms (15 members from 12 institutions)	Direct
Ministry of Water and Environment	Public	Serve as NBC member (1 member)	Direct
Uganda National Bureau of Standards	Public	Serve as NBC member (1 member)	Direct
National Environmental Management Authority	Public	Serve as NBC member (1 member)	Direct
Makerere University	Public	Serve as NBC members (4 members)	Direct
Consumer Education Trust of Uganda	Private	Serve as NBC member (1 member)	Direct
Uganda Consumer Protection Association	Private	Serve as NBC member (1 member)	Direct
Uganda National Farmers Federation	Private	Serve as NBC member (1 member)	Direct
National Agricultural Research Organisation	Public	Serve as NBC member (1 member)	Direct
Ministry of Health	Public	Serve as NBC member (1 member)	Direct
Association for Strengthening Agricultural Research in Eastern and Central Africa	Regional body, public	Serve as NBC member (1 member)	Direct
Ministry of Tourism, Trade and Industry	Public	Serve as NBC member (1 member)	Direct
Advocates Coalition for Development and Environment	Private	Serve as NBC member (1 member); legal expert	Direct
Cotton Development Organisation	Parastatal	Handle all matters pertaining to the country's cotton sector, including release of new varieties	Indirect
Ministry of Agriculture, Animal Husbandry and Fisheries	Public	Authorize seed importation	Direct
Agricultural Productivity Enhancement Program, US Agency for International Development	International	Provide funding and technical assistance	Indirect (operated only until 2008)
Agricultural Biotechnology Support Project, US Agency for International Development	International	Provide technical assistance for CFT implementation	Indirect
Program for Biosafety Systems	International	Provide regulatory advice, capacity building, outreach, and communication; support review and appropriate design of CFTs; develop, with UNCST, CFT guidelines	Indirect
Monsanto	Private	Own technology; design and implement field trials	Indirect

Source: Authors.

Figure 3.6 presents the sequential order of the CFT approval process, shows the most influential stakeholders and actors involved, and identifies bottlenecks in the process. NBC, the Crop Research Institute, and the Ministry of Agriculture are the most influential actors. NBC sets the appropriate biosafety standards and conditions to implement the CFTs and is in charge of their final approval, which requires a consensus among the 15 members of the NBC.

FIGURE 3.6   Net-Map results: Actors and influence in the approval of the confined field trials

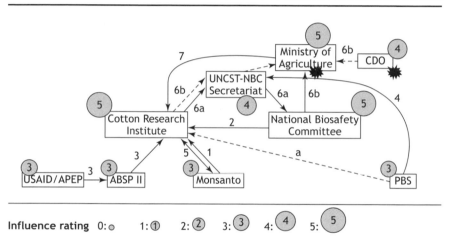

Sources: Developed using Net-Map Toolbox and expert opinions.

Notes: Numbered arrows represent the sequential order of approval for confined field trials; the order of events follows the arrows' directions. Thus, arrow 1 links Monsanto to Cotton Research Institute (CRI) and represents a joint CRI-Monsanto initiative for the implementation of insect-resistant cotton confined field trials. Arrows numbered 6 represent the process by which the seed import permit was issued by the Ministry of Agriculture (MoAg). The size of the circle on each actor's label represents the actor's degree of influence. The ranking goes from 0 (no influence) to 5 (the highest degree of influence). These degrees of influence are based on the subjective assessment of a focus group. ABSP II = Agricultural Biotechnology Support Project II; APEP = Uganda Agricultural Productivity Enhancement Program; CDO = Cotton Development Organisation; NBC = National Biosafety Committee of Uganda; PBS = Program for Biosafety Systems; UNCST = Uganda National Council for Science and Technology.

Each actor plays a specific role in the decisionmaking process. The Ministry of Agriculture has the highest perceived degree of influence, as it is in charge of issuing the seed importation permit necessary to start the CFTs. The Crop Research Institute partnered with Monsanto to implement the CFTs. The Uganda National Council for Science and Technology (UNCST) held the secretariat of the NBC and as such was also among the more influential agents in the decisionmaking process. The Agricultural Biotechnology Support Program II (managed by Cornell University), the Program for Biosafety Systems (PBS), and the Agricultural Productivity Enhancement Program contributed to the review and implementation of CFTs for GM bananas and cotton and to the development of a science-based regulatory framework.[8]

A position in favor of or against GM technologies influences the length of the regulatory process, especially when the position is based not on economic or technical justifications but on political reasons. In this particular case, the approval of CFTs was affected by delays that were mostly bureaucratic but were also the result of dealing with a controversial technology. Reviewing and processing a seed import permit at the Ministry of Agriculture took longer than initially expected. The permission was finally issued in March 2009. However, subsequent applications (for example, for GM cassava materials) have been processed expeditiously.

CDO is the single most important cotton-related institution and as such has decisionmaking power over what seed actually reaches farmers. This institution did not formally participate in the CFT approval process, but it is not hard to imagine that the position and perception of CDO regarding GM cotton would play a crucial role in the commercialization of this technology. Up until now, CDO has promoted organic cotton systems, although it has also shown interest in the adoption of GM cotton as a way to improve cotton productivity. At the same time, CDO has greatly favored the policy of distributing only one cotton variety to all farmers, because, according to CDO's evaluation, this policy has helped guarantee cotton quality in Uganda. All these factors will require the design and implementation of strategies to avoid the mixing of GM and organic seeds, and they may pose challenges to CDO's one-variety policy. It is likely that introducing GM technology will involve other varieties or even hybrids, unless the country inserts those GM genes into their local variety.

---

8 The Agricultural Productivity Enhancement Program stopped operating in the country in 2008.

Approval of the Biosafety Bill

Like the approval of the CFTs, the development of a fully functional biosafety regulatory framework has been a lengthy process involving multiple agents (Table 3.4). A Biotechnology and Biosafety Bill was first drafted in 2003, under a project supported by the Global Environmental Facility of the United Nations Environment Programme, to help Uganda develop and implement its national biosafety framework. The draft bill has been forwarded to cabinet for endorsement and currently remains there. (While the bill was being drafted, a Biotechnology and Biosafety Policy was developed and approved by the Cabinet in 2008.) After cabinet approval, the proposed bill will be submitted to parliament for debate and approval. Following calls to broaden the bill's scope to include biosecurity issues, the Uganda National Academy of Science (UNAS), in collaboration with PBS, spearheaded a process of consultations and studies in 2009 that looked into this question. The UNAS consensus

TABLE 3.4   Actors involved in the design and approval of a biosafety bill

Institution or agency	Type	Role	Influence
Presidential cabinet	Public	Provides approval of bill	Direct
Members of Parliament (MPs)	Public	Provide approval of bill	Direct
Ministry of Finance, Planning and Economic Development (MoFPED)	Public	Provides legal framework for science and technology	Direct
Uganda National Council for Science and Technology (UNCST)	Public	Serves as national biosafety competent authority and provides draft bill for MoFPED	Direct
Global Environmental Facility of the United Nations Environment Programme	International/ donor	Supports development of a biosafety bill	Direct
Program for Biosafety Systems	International	Provides advice to the UNCST and reaches out to the Ministry of Finance and Planning and MPs through targeted forums and information	Indirect
Agricultural Biotechnology Support Project II	International	Supports outreach and provides information on genetically modified organisms and biosafety	Indirect
Uganda Agricultural Productivity Enhancement Program, USAID	International/ donor	Funds UNCST efforts to develop a biosafety framework	Indirect (operated only until 2008)
National Environmental Management Agency	Public	Serves as a focal point for the national Convention on Biological Diversity secretariat	Indirect
Uganda National Academy of Science	Public	Leads a study to advise stakeholders regarding the suggested inclusion of biosecurity issues in the bill	Indirect

Source: Authors.

report concluded that, among other things, the biosafety of agricultural GM crops should be covered through a separate regulatory framework and called for the timely adoption of the Biosafety Bill. However, the bill has yet to be approved. The UNAS study was conducted in parallel with the bill's ongoing approval process.

The process by which the Biosafety Bill has moved from the UNCST to the Ministry of Planning and Finance is mapped in Figure 3.7. This figure presents the steps taken so far and those that must follow to approve the bill in the national parliament. Influence ratings are also illustrated in this figure. As

FIGURE 3.7   Net-Map results: Actors and influence in the approval of a biosafety bill

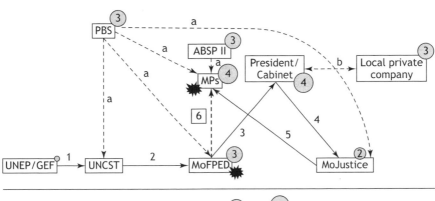

Influence rating   0: ○   1: ①   2: ②   3: ③   4: ④

——▶ Legal process step by step

1 Support project to draft bill
2 Provide draft bill
3 Formulate, gazette (announce) comprehensive draft bill
4 Submit approved draft bill; MPs debate for broader scope of legislation
5 Return formulated bill
6 Submit bill

- - ▶ Ongoing informal links

a = legal and process advice
b = advice

✹ Bottlenecks: Delays

Sources: Developed using Net-Map Toolbox and expert opinions.

Notes: Numbered arrows represent the sequential order of the Biosafety Bill approval process. The size of the circle on each actor's label represents the actor's degree of influence. The ranking goes from 0 (no influence) to 5 (the highest degree of influence). These degrees of influence are based on the subjective assessment of a focus group. Bottlenecks are represented with stars. ABSP II = Agricultural Biotechnology Support Project II; MoFPED = Ministry of Finance, Planning and Economic Development; MoJustice = Ministry of Justice; MPs = members of Parliament; PBS = Program for Biosafety Systems; UNCST = Uganda National Council for Science and Technology; UNEP/GEF = United Nations Environment Programme / Global Environment Facility Coordination.

in the previous case, these values are based on best estimates and expert consultations and would have to be validated with local authorities and actors.

The most influential actor is UNCST, followed by the cabinet and members of parliament. UNCST is the focal point for biosafety and has been responsible for developing the Biosafety Bill since 2003. The bottleneck identified at the time of this exercise was in the Ministry of Finance, Planning and Economic Development, which took a long time to prepare and submit the bill to the cabinet. The parliament may also be another bottleneck, as some stakeholders may not yet be convinced about limiting the scope of this legislation. It is hoped that the UNAS study report will help guide the debate and decisionmaking.

Over the years, PBS, the Agricultural Biotechnology Support Program II, the Global Environmental Facility of the United Nations Environment Programme, and the Agricultural Productivity Enhancment Program have supported—by financial means, technical means, or both—the efforts to develop a biosafety bill. PBS in particular has been very active in giving legal advice to national authorities and promoting forums with the participation of members of parliament.

This exercise helps actors understand the different steps taken for the approval of the Biosafety Bill and also illustrates how the passage of such a bill is a lengthy and complicated process that requires the active participation of multiple actors. Even though the bill was drafted eight years ago and multiple actions have been taken to accelerate this process, delays continue to characterize the approval.

Conclusions and Policy Recommendations

Cotton cultivation in Uganda has a long tradition. Today this commodity, despite its low profitability and numerous production constraints, is still cultivated across the country. Low-input traditional production systems have characterized cotton cultivation. This has created an incentive for using organic production methods, particularly in northern states, and the use of these methods has increased. Therefore, the potential introduction of GM cotton needs to take into account the implications for both conventional and organic production systems.

The cotton value chain in Uganda is vertically integrated: a handful of institutions, such as NARO and CDO, play pivotal roles in research and development and the distribution and commercialization of the country's cottonseed and seed cotton. This vertical integration, on the one hand, can facilitate the dissemination of GM seed but, on the other hand, it can present potential constraints on GM seed introduction. Seed multiplication, distribution, and commercialization could be favored by vertical integration, as well as segregation of organic and conventional cotton. However, the policy of maintaining a single variety can become a problem for the GM cotton industry. Currently, research and development is performed by NARO and, although the organization has been active in selecting varieties, the research and development work needed for breeding a GM variety using local germplasm will demand more resources than the system currently manages.

One of the crucial factors in determining the benefits of a technology is the time that it will take for such technology to reach farmers, as all delays will negatively affect the stream of benefits. If a country has not put in place an adequate regulatory system, the technologies under consideration might take years to be commercially released, might never reach the field, or might become obsolete. In Uganda, the regulatory institutions could be in place, but the delays in these institutions' processing of applications can and are affecting the number of years that it might take for technologies to be released for commercialization.

Instead of relying on educated guesses, as is the case in many ex ante evaluations, our study proposes that, before embarking on the economic evaluation of potential GM technologies, countries analyze and understand their regulatory systems and identify bottlenecks, not only to determine possible delays but also to identify specific bottlenecks that can be addressed by policymakers or decisionmakers. The lessons learned from the analysis of institutions show that complicated processes, such as field trial approvals, can be schematized in a relatively simple graph that can be understood and analyzed. The results can and have been used to work on those bottlenecks and have clarified the potential problems that the approval of these technologies can face. The analysis of the institutional setting also serves to feed the economic analysis in Chapters 5 and 6. In this way, regulatory lags estimated and used in the economic analysis are based on facts rather than on numbers drawn from other countries' experiences.

Chapter 4

Setting the Foundation:
Uncovering Potential Constraints on the Delivery and Adoption of GM Cotton in Uganda

Patricia Zambrano, José Falck-Zepeda, Theresa Sengooba,
John Komen, and Daniela Horna

The government of Uganda has already taken steps to evaluate and eventually commercialize genetically modified (GM) cotton. Uganda has drafted a Biotechnology and Biosafety Bill that is currently under parliamentary review. Meanwhile, the biophysical evaluation of insect-resistant GM cotton is under way, and the two-year cycle of confined field trials has already been completed. However, a better agronomic performance of GM cotton relative to conventional cotton, under confined trials, may prove to be insufficient reason for decisionmakers to approve GM cotton in Uganda. In fact, the competent Ugandan authorities are now interested in evaluating the benefits of GM cotton adoption for cotton farmers and the overall economy. Although these socioeconomic considerations are not part of the Uganda Biosafety Bill, their implementation is under discussion.

An ex ante evaluation needs to make assumptions about different variables that feed into the analytical models. Taking into consideration factors that may affect adoption and delivery of the technology will result in a more robust analysis and will strengthen these assumptions. Examining how current conditions, including institutional capacity, affect adoption and delivery of the technology and how to incorporate these conditions into the economic analysis allows researchers to evaluate the impact of GM technology. Practitioners will plainly benefit if they first assess the institutional requirements needed to achieve impact. The objective of this chapter is therefore to identify potential constraints to GM cotton delivery and adoption and derive from these identifications insights for the economic evaluation in Chapters 5–7.

Factors Influencing Technology Adoption

Feder and Umali (1993, 216) define innovation as

> a technological factor that changes the production function and regarding which there exists some uncertainty, whether perceived or objective (or both). The uncertainty diminishes over time through the acquisition of experience and information, and the production function itself may change as adopters become more efficient in the application of the technology. Adoption can be seen as the process of incorporating the new technology into agricultural practices over time. Under this broad definition of adoption, we include delivery of the technology.

Feder, Just, and Zilberman (1985) define three distinct types of technology adoption processes: individual or aggregate, singular or bundled, and divisible or individual.

The factors most often cited to explain the variability seen in agricultural technology adoption and its diffusion patterns are those described by Feder, Just, and Zilberman (1985). A partial list of explanatory factors would include human and financial capital, farm size, risk impact, attitudes toward and perceptions of risk, family and hired labor, credit constraints, land and other property ownership and formal and informal tenure, and access to commodity markets. These factors are not discrete and are not mutually exclusive, as they often interact with one another in practice. These factors may influence significantly the adoption and diffusion process of new technologies. In fact, the objective of most adoption studies is to examine how each of these variables has affected adoption with the intention of improving the potential release of future technologies.

The technology adoption literature provides guidance on the types of factors that may be relevant in the adoption process. However, we tend to agree with Alston, Norton, and Pardey (1995) that if the purpose of the assessment is the measurement of research impacts from using a new technology, then it is quite difficult to describe adoption in great detail for those technologies that have not yet been adopted. According to Alston, Norton, and Pardey (1995), the best that can be done is to make an informed guess about the likely effects of these technologies on yields or some specific cost item while maintaining other things constant that in fact are likely to be changed too.

What we can do instead is describe the institutional setting before the introduction of the technology and identify potential bottlenecks to technology delivery and adoption.

Institutions and GM Technology Adoption

Institutions, culture, market structure, and the underlying learning process are specific factors that shape the adoption and diffusion of new seed varieties (Lybbert and Bell 2010). The literature has shown that the sustainable adoption of agricultural innovations, specifically GM technologies, has to be supported by adequate institutions that can provide reliable and timely information and services to farmers (Gouse et al. 2005; Smale et al. 2009; Tripp 2009). Comparable to that on any other agricultural innovation, the economic literature on GM cotton underlines the importance of institutions, which can facilitate or hamper adoption, particularly in the context of small-scale farmers.

In addition to possible constraints that also affect the adoption of conventional varieties or hybrids, the commercialization of GM seeds will be preceded by a biosafety regulatory process that is far more stringent than those regulations that are in place for conventional seeds. Countries are required to have a working biosafety regulatory framework in place to be able to evaluate and assess the environmental and food safety risks of the technology. This will require the existence of a competent national biosafety committee that has to be given the legal authority to make these regulatory decisions.

Reaching this point has taken and continues to take a substantial amount of time and legislative effort for many developing countries. Aside from this legal framework, the fact that GM seeds are substantially more expensive than conventional seeds—in Burkina Faso, for example, they are more than five times as expensive as conventional seeds—requires that farmers, particularly resource-poor farmers, have access to adequate credit. Once the crop has been planted, farmers require timely and accurate information to be able to secure the full potential of the technology. Farmers using a technology—such as insect-resistant (Bt) or herbicide-tolerant cotton—need to time applications of insecticides, herbicides, or both properly, as the technology requires specific pest and weed management techniques that are different from those used for conventional seeds.[1] GM seed is mainly a risk-reducing technology, and its efficacy can be significantly affected by stochastic environmental conditions (Lybbert and Bell 2010). Thus, appropriate extension support can be key to a successful adoption process. In addition, because the technology also demands planting the recommended refuge areas for the purpose of

---

1 Bt cotton incorporates insect protection though the inclusion of a gene that expresses a special protein. This protein is toxic to specific lepidopteran species but not to humans, other mammals, or most insects. The origin of the insecticidal protein is the soil bacteria *Bacillus thuringiensis*, hence the name Bt cotton used in the literature.

controlling for insect resistance, its actual enforcement requires an important institutional effort.

Aside from these points, there is also an important difference between publicly developed conventional seeds and GM proprietary technologies. Countries that wish to access these proprietary technologies—Bt and herbicide-tolerant crops are both dominated by the private sector—need to have in place some kind of plant variety protection that guarantees not only a return to investments for the developers of the technologies but also stewardship. All these factors demand a careful analysis of the limitations that the deployment of the technology might face and that can be properly accounted for in the ex ante assessment.

Adoption of GM Technologies: Experiences of Other Countries

A relatively large body of literature documenting and assessing the impact of GM crops in developing economies has been accumulating over the past few years (Qaim 2009; Smale et al. 2009). In particular, the success of Bt cotton commercialization in China has been the subject of many detailed and robust studies by Huang and colleagues (2001, 2002a–c, 2003, 2004). The adoption of Bt cotton in South Africa, and specifically by 100 farmers located in the Makhathini flats, has been the subject of at least 15 peer-reviewed papers since the cotton's introduction in 1996 (Smale et al. 2009). In India, Bt cotton has rapidly spread to most production areas, reaching an adoption rate of 87 percent in just a few years and doubling cotton production. The assessment of this adoption has been done by Bennett et al. (2004, 2006); Bennett, Ismael, and Morse (2005); and Morse, Bennett, and Ismael (2005a,b, 2007), among others. Less-studied cases are those of Bt cotton in Mexico, Argentina, and Colombia.

China successfully adopted Bt cotton and benefited from a very strong research program that developed its own high-yielding, low-cost Bt hybrids, which have spread among most of the cotton production areas. Nevertheless, despite the indisputable fact that Bt cotton has produced higher yields, the reduction in insecticide use has not been fully realized because of the lack of information in farmers' hands (Pemsl, Waibel, and Gutierrez 2005; Xu et al. 2008) and the nontransparent seed markets.

In the case of South Africa, the initial success of Bt cotton was limited by weaknesses in the institutional framework under which it was initially released (Gouse et al. 2005; CDO 2006). Vunisa, the company that introduced the technology in South Africa, implemented a semiformal credit arrangement with cotton farmers that was very successful during the first few years following introduction but was not sustainable in the long run. In fact, when a competitor appeared in the market, undercutting Vunisa's profits, the company could not continue issuing credit to defaulting farmers who were selling their cotton to Vunisa's competitor. This limited access to credit was compounded by abiotic stresses (mainly drought) and scarce availability of irrigation and agricultural inputs.

A detailed review of the different studies on Bt cotton in India (Smale et al. 2009) reveals that the national averages of benefits that the studies present to document the success of Bt cotton in India mask a high variability among the heterogeneous regions in the country. Among the reasons explaining this variability is that farmers have different farming practices and limited access to extension services. The seed market is characterized by the presence of probably hundreds of seeds that change yearly, giving farmers little chance to learn the characteristics of these seeds.

In Colombia, cotton growing is organized into regional associations that provide farmers with an array of services from credit to inputs, including seed and even some limited extension services (Zambrano et al. 2009). Given a small market, as in the case of Mexico, the GM cotton commercialized in the country is planted with imported seed that is locally tested. Despite the initial benefits documented for GM cotton, there is evidence that farmers make a significant number of improper herbicide applications. The studies carried out by a national association suggest that the main constraint on reaching the technology's potential is the lack of information in farmers' hands, from limited or incorrect knowledge of what the technology is targeting to poor practices in the management of the crop.

Something similar might be happening in Burkina Faso. After the impressive adoption of Bt cotton during the first and second year following its introduction, there have been complaints related to seed performance. Although some of the responsibility for this low performance has been put on poor agronomic practices, farmers blame the technology. The debates over the reason for low performance appear to be supporting changes in regulations that could eventually lead to the enforcement of strict liability in the country. Vitale et al. (2010) also document problems with the actual number of pesticide applications. The authors argue that some farmers believed the use of Bt seed meant no applications were needed, seriously affecting the performance of the crop. Although there has been no evaluation of the extent of the complaints about seed performance, the circulation of misinformation in Burkina Faso highlights the importance of having the right information flowing to and from farmers.

The country cases described point to the critical need to address and understand the possible barriers to the successful deployment of GM crops. The experiences of other countries with the adoption of GM cotton suggest that its success is linked to the institutions that are behind the organization of cotton production. The characteristics and strengths of institutions that provide timely and accurate information, seeds and other inputs, and credit resources to farmers can enhance the successful adoption of GM cotton. As with other technologies, such as high-yielding hybrids, an efficient and successful adoption of GM seeds requires a change in agronomic practices for farmers to reap full benefits from the technology. Farmers need to have timely access to these seeds and to be capable of choosing—and having the option to choose—between GM and non-GM varieties. Credit and input channels have to be in place to allow cotton farmers to access seeds and other inputs. Another critical point is the price of the technology and government or local agents' ability either to negotiate with technology owners to deliver the GM seed at competitive prices or to provide sufficient credit to buy significantly more expensive seeds. In Burkina Faso, the parastatal cotton company extends credit to most cotton farmers, thus enabling farmers to pay for Bt seeds that are five times more expensive than conventional seed.

GM Adoption in Uganda

To evaluate whether the current organization of the cotton production industry is an enabling environment for the successful deployment of GM cotton in Uganda, our study first identified the different limitations in the cotton sector that could affect the commercialization of the technology. We identified the actors through both a review of the available literature related to the institutional setting of cotton production in Uganda and face-to-face interviews with experts and representatives of different institutions. Rather than creating a list of institutions and their formal structures and functions, we focused on assessing whether the institutions in Uganda provide services and information identified in previous studies as essential to the successful adoption of the technology.

The study seeks to answer three sets of questions. First, which institutions are involved in the distribution of seeds and how would the system adapt the distribution of both GM and non-GM seed to resource-poor small-scale farmers? Second, which institutions are powerful (at least in terms of monetary resources), and where are the resources to finance the successful adoption of GM technology? Third, given the key role that information plays in the adoption of the technology, which institutions or agents are providing this information and how much information reaches farmers? To answer these questions, the study mapped all relevant agents. The results of this exercise are illustrated in Figure 4.1. As in the case of the regulatory institutions detailed in Chapter 3, the background information and expert assessment involved in the exercise were drawn solely from the literature review and interviews with study collaborators in Uganda mentioned above. The study also used Net-Map (Appendix 5) to collect and structure the data.

Figure 4.1 shows a map with 15 actors and institutions that could potentially play a role in the deployment of GM cotton to farmers. Three different types of links were established among these actors. The first link is seed distribution, the second link is monetary flows among the different actors, and the last is information distributed and received among different agents. The most solid and relevant link among actors is the money flow, which gravitates around the cotton ginneries. Given that the cottonseed market is centrally controlled by the Cotton Development Organisation (CDO), the seed flow is mainly from CDO through the ginneries to farmers. The weakest link

FIGURE 4.1 Current institutional setting affecting the commercialization of GM cotton

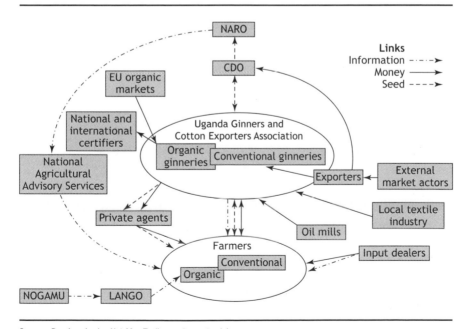

Sources: Developed using Net-Map Toolbox and expert opinions.
Note: CDO = Cotton Development Organisation; GM = genetically modified; LANGO = Lango cooperative union; NARO = National Agricultural Research Organisation; NOGAMU = National Organic Agricultural Movement of Uganda.

is information. Although different agents link to farmers with information, the number of agents effectively providing this service is very limited. The assessments of each of these links' role in the deployment of GM cotton are detailed below.

Seeds

The vertical integration of the seed value chain, as described in Chapter 3, creates limitations in the way seed—specifically, GM seed—can be produced and distributed. The system already has limited capacity to produce conventional seed in the quantity and quality needed. If transgenic seed enters the market, its entry will add to the current constraints in the system and will require policy decisions to guarantee its production and effective distribution. Cotton cultivation in Uganda is predominantly done by small-scale farmers scattered around the country. CDO has a relatively successful system of seed distribution, but it would still need to be adapted to the distribution of GM seeds. The main limitation of this system is that CDO currently distributes a unique seed that is used by both conventional and organic farmers (see Figure 3.5 and Chapter 3). This policy would need to be revised if GM cotton is approved for commercialization in Uganda. CDO, as the current governing body for the distribution of cottonseed, will need to assess several possible scenarios:

- *Continue with the one-variety policy.* If CDO wants to continue enforcing the one-variety policy and the country embarks on the commercialization of GM cotton, then the only variety made available to farmers will be a GM variety. Under these circumstances, organic farmers would be left without seeds, as the European market—where most organic exports go—would not accept cotton grown with a GM variety, even if no pesticides are applied. This is an unlikely scenario for two main reasons. First, the organic cotton sector is perceived as an important one, particularly for the income it generates from European markets. It is not likely that the country will opt to give up this market to allow commercialization of GM cotton. Second, it poses questions on what variety would be commercialized. Would it be a new GM cotton variety, or is it expected that the current cotton variety would be transformed to include the desirable GM traits? The first option, a new imported variety, is what many small-scale countries have done, but it implies that the perceived benefits of CDO's one-variety policy would be eliminated, unless the new variety is superior to the one currently in the market. The second option, transforming the current variety, is costly both in terms of financial resources and time invested.

- *Allow the distribution of two distinct varieties.* Under this scenario, one variety would be the local one (BPA 52) that both organic and conventional farmers currently use. The second variety would be an imported GM variety that would be available to farmers only in specific areas. If the policy were to preserve the organic sector, Uganda would probably need to establish some kind of zoning to facilitate segregation. It is likely that the GM variety made available to farmers would be an imported one, as this has been the experience so far in all small adopting markets like the one in Uganda. South Africa, Colombia, and Mexico, among others, have used imported GM varieties, as the size of the cotton market does not justify the necessary investments to insert the technology into the national germplasm. India and China have gone through the technology-insertion process because their markets are quite sizeable and they have the local institutional capacity and resources to accomplish this task. In Burkina Faso, where cotton is the main agricultural activity, commercialized Bt cotton was also done using local varieties.

- *Use two isogenic varieties.* Under this scenario, CDO continues with BPA 52 but requires the insertion of the technology into this variety, so the organization can continue to assert control over the fiber quality. This scenario would require additional time and human and financial resources for research and development. The development of its own isogenic variety could be a great asset to Uganda if the country can capture seed markets outside its borders.

- *Open the seed market.* This scenario implies a change in the vertical integration of the cotton chain and potentially weakens the control that CDO currently has over the type of seed to be commercialized. However, it might be the best alternative for the country if the breeding process, selection process, or both is done using varieties selected and adapted to Ugandan conditions. This scenario might not be the most favored in the current institutional setting, but it could favor the short- and long-term performance of the crop. The use of a single variety, regardless of its quality or adaptability, poses sustainability threats to cotton production, because lack of diversity increases environmental stresses and diseases (Smale, Bellon, and Pingali 1998). Adopting countries in Latin America have in fact imported GM cotton varieties that have been properly evaluated for domestic conditions and have successfully commercialized such varieties. It is a less costly path, both financially and time-wise, and one worth considering for adopting countries with a relatively small market size.

These scenarios assume that CDO or another government agency will be able to negotiate a technology price that will guarantee that farmers get the seed at a price that is beneficial for all stakeholders. Uganda is a small market for Monsanto or any other biotechnology company, but it can serve as a critical showcase for GM products in East Africa, which can give the national authorities some negotiation leverage. Developing a Monsanto/National Agricultural Research Organisation variety that can be marketed in other countries could be embedded in these negotiations.

Information

Timely information about the characteristics and specifics of a new technology is a key element in the successful adoption of the technology. Information is particularly relevant in the case of GM technologies, as in many cases they require changes in agronomic practices (Tripp 2009). Farmers need to know how the technology works and to understand the specifics of which constraints this technology is targeting. Both Bt and herbicide-tolerant cotton require farmers to modify their practices with respect to pesticide use and seed management. To realize the full potential of this technology, farmers need to know basic pest epidemiology, so that they know when and where it is—or is not—appropriate to apply insecticides. Farmers need timely information about the specifics of how the GM technology works (Pemsl, Waibel, and Gutierrez 2005). For instance, farmers must be aware that even if bollworms are in the plot, it is not always necessary to apply insecticides and that herbicides can be applied directly to the plant but only at specific times. Results from Vitale et al. (2010) for the case of Burkina Faso confirm this assessment.

How this detailed information is going to get into the hands of farmers needs to be determined. Current information channels in Uganda are few and appear not to be sufficient to guarantee that this information will flow in a timely manner. Figure 4.1 shows the different information links among the actors, which at first glance appear to be abundant. Nevertheless, the information links from the National Agricultural Research Organisation to the National Agricultural Advisory Services, and from the latter to farmers, are quite limited.

As described in Chapter 3, after the collapse of the zoning system in 2007, there was no longer an incentive for ginneries to provide extension services for

cotton producers in their zones. Currently, the only ginneries that continue
to give extension services are a few big ones that can guarantee that the ben-
eficiary farmers will sell the harvest back to them. In an effort to compensate
for the lack of extension services, CDO has established a limited extension
scheme of hiring an extension agent for each production district. The scheme
has been designed to reach farmers at a regional level, encourage some community mobilization, and enable farmers to organize themselves. Given the size of
these districts and the number of cotton farmers in each district, the scheme's
capacity is very limited. The so-called extension agents are short-term employ-
ees who have little incentive or opportunity to develop close relations with
farmers. Because of this, the agents have no direct influence over farmers' agronomic practices, which would be desirable for the adoption of GM cotton.

With regard to organic farming, there are also important information
needs, particularly if the government is interested in preserving this system. For organic farmers, the main source of information is services provided
directly through organic ginneries, such as Dunavant, and some cooperatives, such as Lango. How good and extensive this information is has yet to be
assessed. The potential impact of the deployment of GM cotton on organic
operations and the need to design strategies for a proper segregation are two
examples of information gaps that need to be addressed.

Money and Financial Resources

The development of GM seed for Uganda represents a significant investment. It is quite likely that GM seed would not be delivered free of charge, and
either the government, CDO, the ginneries, farmers, or all four would need to
cover the cost of the technology, as has been the case in most adopting countries. The analysis of the flow of financial resources in the systems can provide
insights into what sorts of incentive mechanisms need to be in place to pay for
the technology and stimulate adoption. It is possible that the government of
Uganda would assume part of the cost, but it can also be expected that farmers
would pay for the technology. As mentioned earlier, seed is currently perceived
as a free good, so making farmers pay a technology fee would need to result
in more than just extension and experimental plots. For resource-poor farmers, the big incentive to adopt a technology is often a guaranteed profit, especially for a cash crop such as cotton. Few farmers want to invest in a new and
complex seed technology of a low-profit–high-risk crop. It makes sense, therefore, to evaluate the potential profitability of GM cotton and its contribution
to risk reduction. If a technology increases expected profits by eliminating or
reducing risk, risk-averse farmers will most likely adopt it even if they have to

pay. Ideally, one could be able to derive ex ante willingness-to-pay estimates (such as the demand for GM seed) for different regions under various assumptions about risk preferences. To take this step, however, it would be preferable to have more detailed information about the producers in the different regions. This information could be extremely useful for implementing a price-discrimination strategy.

Although in Figure 4.1 producers receive cash inflows from different agents, these are resources they receive after harvest time. The fact is that they have limited or no access to credit or resources to finance their cotton operations. This situation may hamper adoption of new technologies and seed varieties, particularly if these seeds are provided at a cost. The economic analysis may show that the technology is affordable but requires farmers to have access to credit to finance the initial investment. The few credit sources available to cotton farmers have been limited. Individual ginneries experimented with input credit to farmers without much success, as it was extremely difficult to recover credit from farmers. The Uganda Ginners and Cotton Exporters Association also had to abandon its efforts to provide inputs to farmers for similar reasons (Gergely and Poulton 2009). Given the limited ability of farmers to repay their credit, it is hard to envision how this service could be provided without having in place some mechanism that would guarantee its recovery.

That most money arrows in Figure 4.1 are directed to and from ginneries is an indicator of the concentration of economic power in the ginneries. The leading economic interest that ginneries have in GM technology is likely to influence the technology's delivery. Ginners generate most of their income from selling conventional cotton in external markets, although they also benefit from a smaller but seemingly profitable organic European market. A source of income for CDO comes in the form of export permits all ginneries are required to have to export their cotton lint. If GM cotton were to be commercialized, ginneries—particularly those ginneries that are certified organic and the ones that move back and forth from conventional to organic—will need to monitor the movement of cotton. Such commercialization has benefits and costs for ginneries. On the one hand, ginneries will likely welcome the opportunity to increase cotton production, as the industry's capacity utilization is just 20 percent. On the other hand, the commercialization of GM cotton will mean the end of some ginneries' movement, from one season to the other, between handling conventional and organic cotton. The economic impact on the cost of cotton processing that this change will produce will need to be assessed.

Conclusions and Policy Recommendations

Successful diffusion of transgenic cotton in developing countries relies on functional public institutions. To be specific, it relies on the clear support of the state, sufficient resources to guarantee the deployment of the technology, a dynamic seed-distribution research sector, and timely and accurate information in the hands of farmers. Discussion among relevant actors and decision-makers is needed regarding the current institutional setting.

Our analysis shows that the government of Uganda, including CDO, needs to address specific questions regarding seed production and distribution, access to credit, and access to information to support GM technology delivery and adoption by farmers:

- More discussion is needed about the type and price of GM seed that will be distributed to farmers. Alternatives regarding which variety or varieties will be made available must be assessed. In Uganda, cottonseed is currently freely distributed to farmers, but further analysis is necessary to assess how realistic and sustainable this policy is for a proprietary technology. Even if hypothetically the technology becomes freely available, mechanisms must be in place to guarantee that the seed will not be multiplied illegally and also that good stewardship practices are in place, as required by technology owners. This decision will critically affect the research and development lag and would have a direct impact on the final cost of introducing GM cotton in Uganda.

- The current weak information links to farmers need to be strengthened. The adoption of GM seed will bring changes to the agronomic management of cotton, and farmers have to understand the technology and its significance with respect to reduced use of insecticides, increased use of herbicides, or both. If better communication does not occur, then initial adoption might be unsustainable. The public sector championed by CDO needs to involve other actors in the delivery of the technology and guarantee that extension agents, such as the National Agricultural Advisory Services, receive appropriate training.

- The market would definitively influence the decision on the type of GM variety. In this context, it is particularly important to discuss how to introduce the necessary adjustments to guarantee that the commercialization of GM cotton will have a beneficial impact on the livelihoods of resource-poor farmers. How to address the limited availability of credit to farmers should also be part of the discussion.

Chapter 5

Assessing Genetically Modified Cotton's Economic Impact on Farmers

Daniela Horna, José Falck-Zepeda, and Miriam Kyotalimye

I n Uganda, cotton has been characterized as a crop with relatively low profitability, mostly due to low productivity (Baffes 2009), but also because it
is affected by fluctuations in cotton's world price. Studies done by APSEC
(1998, 2001) ranked cotton as the lowest in profitability among the main
competing crops on the global market. Despite cotton's low profitability,
farmers continue to plant it. The most-often-cited reason for continued cotton production is a lack of productive alternatives that can generate cash for
smallholders and larger farmers during the period cotton is planted. The cer-
tainty that cotton producers will have a buyer at the end of the season is probably another strong argument for cotton cultivation: ginneries usually distribute
seed and inputs and in turn demand rights over the seed cotton harvest at the
end of the cropping season.

In this chapter, the following question is addressed: would the adoption
of genetically modified (GM) cotton make farmers better off? The methods
and tools chosen to address this question had to be adjusted to the Uganda
context, where biosafety regulations are in the process of approval and the
inclusion of socioeconomic considerations is expected to contribute to
decisionmaking. To produce meaningful results within a limited time and a
restricted budget is challenging. It is also challenging to produce these results
for a technology that has not been tested in the country. That GM cotton is
a technology already approved in about 10 countries, including South Africa
and Burkina Faso, does make the task easier, however.

In this study, the primary data comes from the household survey implemented in the two main cotton-producing districts in Uganda: Kasese and
Lira. Secondary sources are also used, especially to support assumptions that
the selected methodological approach demands. The study evaluates yield
performance by using a production function to understand factors determining productivity. Partial budgets are used to evaluate the profitability of cotton
production at the farm level and to compare conventional production with
hypothetical GM crop scenarios. The study adds stochastic simulations to the

partial budgets to account for the effects of risk and uncertainty related to cotton production and its profitability. Although the sampling framework for the collection of primary information does not permit the extension of findings to the country level, this exercise does allow a more detailed understanding of what the impact of GM cotton adoption could be on farmers in the most important cotton districts in Uganda.

Measuring Ex Ante Impact of GM Technologies on Farmers

Although an economic literature exists on the impact of transgenic crops on farmers' welfare, most of these evaluations are ex post (Smale, Niane, and Zambrano 2010). Two main approaches have been used to measure this ex post impact: (1) examining partial budgets that compare net profits from adopters and nonadopters and (2) using a statistical approach in an economics framework, such as a production function or a random utility framework. Even though the second approach allows for a more rigorous test of ex post impact (Smale et al. 2009), the use of econometric tools is more limited in ex ante evaluations, especially in those cases where there is no experimental or trial information that can be analyzed.

There are, however, a few examples of studies that use econometric tools to evaluate farmers' preferences and thus potential adoption of GM crops. Edmeades and Smale (2006) predict farmer demand for disease- and pest-resistant bananas in the East African highlands using a trait-based model and survey data that detail cultivar attributes and the characteristics of farmers, households, and markets. Kolady and Lesser (2006, 2008a,b, 2009) and Krishna and Qaim (2007) explore ex ante the potential adoption and impact of insect-resistant eggplant in India using contingent valuation, production functions, partial budgets, and sensitivity analysis to assess the potential benefits of the adoption of open-pollinated varieties versus hybrids. Birol, Villalba, and Smale (2008) and Kikulwe et al. (2009) examine farmers' preferences and use a latent class model to identify the characteristics of future GM seed adopters. In the first case, the authors characterize farmers and their need for compensation if transgenic maize were introduced to Mexico. In the second case, the authors assess the case of transgenic banana in Uganda.

Although these studies are very important resources for GM crop evaluation, the research question they address—how farmers' preferences affect GM crop adoption—is slightly different from the one addressed in this chapter: how profitable is GM crop adoption? If decisionmakers in a country opt to include socioeconomic considerations in biosafety regulation approval, they would be interested to know whether the adoption of GM crops results in a profitable business. With this objective in mind, the use of partial budgets at the farm level is the preferred option. Partial budget analysis is a simple tool that can assist in understanding cotton profitability and identifying some general constraints on profits. This tool is particularly useful for re-creating farmers' actual conditions and simulating counterfactual scenarios. A common criticism of partial budgets, however, is that they are only snapshots of reality. The use of stochastic simulations in addition to partial budget analyses allows for a more dynamic analysis that can better represent farmers' conditions. This representation of farmer conditions also distinguishes this study from early studies of GM varieties that were only ex ante studies or field experiments (Qaim and Zilberman 2003).

Methodology

To assess the potential impact on farmers of GM seed introduction, we first analyzed the effect of risk and uncertainty on the current performance and profitability of cotton production. We consider this step necessary not only to find out what the main yield determinants are but also to gauge the use of chemical inputs and their effectiveness in controlling the target constraints (bollworm and weeds). This information is crucial for developing scenarios and simulations. As much as possible, we use primary data to determine our assumptions about GM technology's performance. Table 5.1 presents the methodological steps followed and the data used for this assessment.

Production Function

The use of a production function to evaluate the performance of a technology ex ante is possible only if there is at least trial data available. However, this is not the case for GM cotton in Uganda, as the GM technology has not yet been deployed. Nevertheless, an assessment of cotton production can help evaluate the main factors determining crop performance and determine the possible effects of introducing GM seed (as done by Horna et al. 2008).

Our study uses a production function with a damage-control framework to correctly account for the effect on yield of inputs that facilitate growth and control damage. The damage-control framework (Lichtenberg and Zilberman 1986) has been widely used to measure the ex post impact of growing Bt cotton (Huang et al. 2002a,b; Qaim and de Janvry 2005; Shankar and Thirtle

TABLE 5.1   Methodological steps

Step	Method	Data source
1. Evaluate current yield determinants of cotton	Production function using a damage-control framework	Survey data
2. Estimate current profitability, risk, and uncertainty of cotton production systems (conventional and organic)	Partial budget analysis Stochastic simulations of conventional and organic cotton production	Survey data and assumptions about GM technology performance Simulated distributions based on survey data
3. Forecast cotton profitability with GM seed introduction	Stochastic simulations of cotton production using GM seed (Bt and HT)	Simulated distributions based on (i) survey data and (ii) assumptions about GM technology performance
4. Analyze GM seed introduction's impact on farmers, given different payment arrangements for the technology fee	Stochastic simulations of cotton production using GM seed (Bt and HT)	Simulated distributions based on (i) survey data, (ii) assumptions about GM technology performance, and (iii) assumptions about potential technology fees
5. Evaluate the effect of complementary inputs: labor, fertilizer, and pesticides	Sensitivity analysis	Assumptions based on best available information

Source: Authors.
Note: Bt = insect resistant; GM = genetically modified; HT = herbicide tolerant.

2005). As explained by previous studies' authors, the damage-control framework considers that agricultural inputs, such as insecticides and pesticides, are not yield enhancing but loss abating. The damage-control effect is defined as the proportion of the destructive capacity of the damaging agent that is eliminated by applying a certain amount of a control input. Control inputs can be pesticides, additional labor, cultural practices, a crop variety (including a GM variety), or any other input that the farmer uses to mitigate the impact of pests and diseases. A standard production function in a damage-control framework is specified as

$$Y = F[\mathbf{Z}, G(\mathbf{X})], \tag{5.1}$$

where Y is the crop output, the vector $\mathbf{Z}$ represents directly productive inputs, and the vector $\mathbf{X}$ represents the control inputs (Lichtenberg and Zilberman 1986). The damage-control function $G(\mathbf{X})$ takes values in the interval $[0, 1]$. If there is no control of the damage, $G(\mathbf{X}) = 0$ and $Y = F[\mathbf{Z}, 0]$; if there is complete control of the damage, $G(\mathbf{X}) = 1$ and $Y = F[\mathbf{Z}, 1]$. In this equation, even though the function $G(\mathbf{X})$ is unobservable, the use of control agents $\mathbf{X}$ can be directly observed and measured.

For a flexible and robust estimation of cotton production in Uganda, this study used a quadratic production function with a logistic abatement function:

$$Y = (\alpha + \sum_i \beta_i \mathbf{Z}_i + \sum_i \sum_j \phi_{ij} \mathbf{Z}_i \mathbf{Z}_j + \gamma \mathbf{H} + \varepsilon) \times ([1 + \exp(\mu - \sigma \mathbf{X})]^{-1}), \quad (5.2)$$

where $\mathbf{H}$ stands for household characteristics, and $\mathbf{Z}_i \mathbf{Z}_j$ represents the interaction effect of productive inputs. For the estimation of this function, the study used the nonlinear least-squares procedure. Weather and soil characteristics can also influence yield.

The damage-control framework can account for the yield effect and also for the pesticide-reduction effect. The yield effect can take different forms that can be grouped into two broad categories. In the first category, a yield effect can be either a damage-control effect or the efficiency of the technology controlling the constraint. In the second category, a yield effect can be either the genotypic advantage of the GM variety or a pure yield effect. Qaim and Zilberman (2003) used the damage-control framework to evaluate the impact of GM cotton varieties in India and concluded that in countries where pest damage is significant and pesticides are not effectively used, the adoption of GM varieties would increase yield mainly by controlling damage but would have less impact on pesticide use. In countries with high pesticide use, the damage-control effect would be small, but the technology would help lower pesticide use and thus reduce health costs. Because the use of complementary inputs is low in Uganda, we expect to have a high damage-control effect and no pesticide effect.

Partial Budgets

To develop partial budgets, we followed the comprehensive guide published by the International Center for Maize and Wheat Improvement (CIMMYT 1988). As explained in the guide, a partial budget is a method used to organize and present benefits and costs of alternative treatments.[1] Reported use by farmers of such inputs as land, labor, insecticides, and herbicides is converted to values per hectare. Cottonseed is distributed free of charge and thus has zero cost for producers. We used the opportunity costs of land, equivalent to the average rent in the district, to account for land costs. Total family labor

---

1 It is typical to include total benefits but the costs only of inputs that vary with treatments. Because we are comparing not only alternative treatments but also production systems, we include in the partial budgets not only the costs of inputs that vary with the treatments but also the costs of inputs that vary with the production system.

costs are estimated using average wages paid to hired labor. As stated by Horna et al. (2008), this assumption is reasonable, because labor markets in selected districts are active and cotton is a commercial crop. Unfortunately, family labor was reported in the survey only as the total number of hours invested in cotton production during the production season and was not disaggregated by activity. Detailed information on hired labor and benefit-cost ratios are reported for both scenarios, with and without the inclusion of family labor. Male and female labor days are assigned equal costs, as there is no evidence available to justify valuing them differently.

To develop partial budgets for GM scenarios, yield losses due to targeted constraints are derived from the elicited yields, following the same steps as in Horna et al. (2008):

$$E(Y_{\text{loss}}) = \frac{\left[E(Y_{c=0}) - E(Y_{i,c=1})\right]}{E(Y_{c=0})}, \tag{5.3}$$

where $E(Y_{\text{loss}})$ is the expected yield loss ratio, $E(Y_{c=0})$ is the expected yield without the constraint, $E(Y_{c=1})$ is the expected yield with the constraint, and i indicates use of insecticide (1 if farmers use insecticide and 0 otherwise). Given expected yield losses, expected damage control with insecticide Y_{abat} is estimated as

$$E(Y_{\text{abat}}) = 1 - E(Y_{\text{loss}}). \tag{5.4}$$

This estimation is a fair approximation of damage control. Actual damage and damage control are variables that are difficult to estimate, whether from field experiments or data obtained from farmers. In the first case, control conditions can hardly replicate farmers' plot conditions. In the latter case, yield losses reported tend to be upward biased, because it is difficult for farmers to isolate the effect of damage from a specific constraint.

Stochastic Simulations

We ran four simulations to account for different production possibilities for farmers. The first simulation depicts an organic production system in which the premium price is replaced by a triangular distribution, where the low premium is 0 percent (when farmers do not get the premium), the mode is 12.5 percent (in cases where the premium is paid only on part of the harvest), and the high premium is 25 percent. This last value is the premium paid to organic farmers (as reported by the National Organic Agricultural Movement of Uganda). The mode value is half of this high premium. These values are used as a conservative approximation, as our records show no difference between the price paid to organic producers and that paid to conventional producers. A factor that may explain this observation is that farmers reported net income received from the ginnery, which may have already discounted transaction costs (transportation or ginnery fees) to a greater degree for organic producers. Moreover, in their evaluation of the organic sector, Ogwang, Sekamatte, and Tindyebwa (2005) reported that organic producers were receiving on average a 13 percent price premium. The second simulation is a hypothetical case in which it is possible to use Bt cottonseed in an organic system. The third and fourth simulations illustrate the case of the adoption of Bt cotton and herbicide-tolerant (HT) cotton, respectively.

From the experience of adopting countries, it is expected that GM seed's price will be higher than current seed prices, although the absolute value of this price varies widely according to the technology provider and its market power. Cost savings associated with the use of GM seed are measured by the reduction in insecticide applications, related labor costs (if any), or both. Assumptions used in the partial budget simulations are summarized in Table 5.2. To account for risk and uncertainty in agricultural production, some parameters were replaced by distributions. The distributions used in the study were based either on a literature review (in the cases of technology fees, the damage-control effect [technology efficiency] and reductions in pesticide and spraying costs) or on primary data collected from farmers (in the cases of yield variability within and across households, yield loss due to constraints, price fluctuations, pesticide use, and spraying).

We used the @Risk software (Palisade Corporation 2012) to generate the distributions. For variables with actual information available (such as season yield, use of inputs, and costs), @Risk estimated candidate distributions and selected the one that best fit the information collected in the survey. For variables that represented farmers' judgments or perceptions—specifically of yield behavior over time—@Risk selected the triangular distribution that best fit the information elicited from farmers at three moments: (1) without the constraint, (2) with the constraint but without insecticide use, and (3) with the constraint and insecticide use. In @Risk, the study drew from the sample distributions of each yield parameter (minimum, maximum, and mode) to identify yield variability both in and across observations. Note that in @Risk's parlance, there are two kinds of variables (distributions): input variables that are predetermined and output variables that are estimated based on input variables.

Best-fit distributions were used for variables that were easy to obtain from farmers: (1) output price, (2) pesticide cost, and (3) spraying cost. In contrast,

TABLE 5.2   Variables and distributions used for partial budget simulations

Variable	Distribution	Assumptions and source
Yield (kilograms per hectare)	@Risk best-fit distribution	Based on information collected from farmers
Yield losses due to bollworm and lack of weeding (percent)	@Risk best-fit distribution	Based on information collected from farmers
Technology efficiency (percent)	Triangular distribution	Values: minimum = 0, mode = 50, and maximum = 100, based on literature for both Bt cotton and HT cotton (Pray et al. 2002; Qaim 2003; Traxler and Godoy-Ávila 2004)
Output price (US$ per kilogram)	@Risk best-fit distribution	Based on information collected from farmers
Seed cost (US$ per kilogram)	Not a distribution	Seed is distributed free of charge; the value assumed was US$0.205 per kilogram; on average, farmers can use 10 kilograms per hectare of seed for planting cotton
Premium price (percent)	Triangular distribution	Values: minimum = 0, mode = 12.5, and maximum = 25; these values represent the increase, for organic producers, over the official price
Technology fee (percent)	Triangular distribution	*Scenario 1:* Values: minimum = 0, mode = 50, and maximum = 100; these values represent the increase over the seed price *Scenario 2:* Range of values found in the literature (Falck-Zepeda, Traxler, and Nelson 2000; Huang et al. 2003, 2004; Bennett et al. 2004), including US$15 per hectare for India, US$32 per hectare for South Africa and China, and US$56 per hectare for the United States
Pesticide cost (US$ per hectare)	@Risk best-fit distribution	Based on information collected from farmers
Reduction rate in pesticide used to control lepidoptera (percent)	Triangular distribution	Values: minimum = 0, mode = 50, and maximum = 100
Herbicide use (US$ per hectare)	@Risk best-fit distribution	Based on information collected from farmers
Increase in herbicide use (percent)	Triangular distribution	Values: minimum = 0, mode = 50, and maximum = 100; these values represent the increase over the average among current herbicide users
Labor for pesticide application (US$ per hectare)	@Risk best-fit distribution	Based on information collected from farmers
Reduction rate in labor used for pesticide application (percent)	Triangular distribution	Values: minimum = 0, mode = 25, and maximum = 50; these values represent the reduction in labor as a result of the reduction in total pesticide applied
Labor for herbicide application (US$ per hectare)	@Risk best-fit distribution	Based on information collected from farmers
Increase in rate of labor used for herbicide application (percent)	Triangular distribution	Values: minimum = 0, mode = 25, and maximum = 50; these values represent the increase over the average, among current herbicide users, of labor used to apply herbicides

Source: Authors.

Note: Bt = insect resistant; HT = herbicide tolerant.

triangular distributions were used to model variables that measure (i) technology efficiency (trait expression), (ii) technology fees, (iii) rates of reduction in pesticide use, (iv) rates of reduction in spraying costs in the case of Bt cotton, and (v) increased rates of herbicide use in the case of HT cotton. An explanation of minimum, mode, and maximum values adopted for all these variables is given in Table 5.2.

The effect on labor of the use of GM varieties depends on the type of material used. The use of Bt cotton varieties can lead to a reduction in the labor employed for chemical applications. However, if the expected yield increase from Bt cotton use is high, then harvesting will require more labor. But the study assumed that yield increases would not necessarily lead to higher labor use, because yield levels are currently rather low in Uganda. The use of HT cotton implies a reduction in the labor used for manual weeding but an increase in the labor used for herbicide application. Not many farmers currently make use of herbicides in Uganda. Of those who do, few apply glyphosate.

The results of the real and simulated partial budgets were compared using first-degree stochastic dominance. Stochastic dominance is a nonparametric approach used to rank competing alternatives, strategies, or policies based on their risk characteristics (see Appendix 6). This approach ranks alternatives into dominating and dominated sets based on stochastic efficiency rules. Stochastic efficiency rules are pairwise comparisons of the estimated cumulative distribution functions derived from observed or simulated data (or both) describing an outcome or action. In most cases, these distribution functions tend to intersect. Thus, additional and more restrictive assumptions are needed to allow the ranking.

Technology-Fee Scenarios

The technology fee is a sensitive issue because GM-seed price affects adoption. This study developed two contrasting technology-fee scenarios to evaluate its effect on farmers' welfare.

In the first scenario, the technology fee is expressed as a percentage increase of the assumed seed price.[2] The study used a triangular distribution

---

2 Currently farmers do not pay for the seed, as it is assumed that it is distributed free of charge, so there is no information about the price of the cottonseed. The study therefore uses the nominal farmgate price charged by farmers for their seed cotton (Tschirley, Poulton, and Labaste 2009). As with other crops, the price of the seed is expected to be higher than the farmgate price of the commodity; this is, however, the best approximation of the seed's real value. Also, note that ginneries select the best cottonseed after delinting and then give it back to the Cotton Development Organisation.

in which the low percentage increase over the value of the conventional seed was 0 percent, the mode was 50 percent, and the high was 100 percent. Because the assumed price of the conventional seed was zero, the market price for the GM seed, including the technology fee, ended up being still much lower than real GM seed prices charged in developing countries that have already adopted the technology. However, this scenario was useful in understanding what the benefits would be to farmers with a technology fee subsidized by the public sector.

In the second scenario, the technology fee is expressed as a triangular distribution in which the low ($15 per hectare),[3] mode ($32 per hectare), and high ($56 per hectare) values are based on the final values charged for GM cottonseed in other countries. This scenario represents the extreme case in which the technology fee is fully paid by farmers.

Characterizing Cotton Producers

Farm and household characteristics have an effect on the adoption of GM cotton. Analysis of GM cotton adoption highlights the importance of farm size, labor availability, access to credit and production inputs, and risk preferences. These and other key variables have a major influence on the adoption of agricultural technologies, especially high-yielding crop varieties (Feder, Just, and Zilberman 1985). The crucial farm household characteristics and production variables are summarized by district in Table 5.3.

The statistics in this table describe well the situation of cotton production across districts in Uganda: small plots, low input use, and low productivity.[4] As mentioned earlier, Lira and Kasese are important cotton-producing districts. Lira is the district where organic production has expanded faster. Kasese is among the most important cottonseed-producing districts, being the largest producer in 2006. The higher concentration of ginneries in this district is probably explained by the favorable conditions for cotton production. Although our sampling strategy was strongly linked to the selection of sites for the implementation of the confined field trials and therefore might suffer from bias, the information in Table 5.3 is still similar to the conditions in other cotton-producing districts in Uganda. Note that, although our sample's mean seed cotton yield of 953.48 kilograms per hectare is higher than the

---

3   All dollars are U.S. dollars in the chapter.

4   The mean size of cotton plots in Lira (0.44 hectares) is comparable to the average plot size of 0.41 hectares reported by UBOS (2007) (which used information from UHNS for 2005/06).

TABLE 5.3   Descriptive statistics for farm households by district

Variable	Total sample (N = 151) Mean	Standard error	Lira (N = 35) Mean	Standard error	Kasese (N = 116) Mean	Standard error	t-test	P-value
Gender of household head (female = 1)	0.09	0.02	0.03	0.03	0.11	0.03		
Control of plot (female = 1)	0.46	0.08	0.29	0.13	0.51	0.09		
Age of household head (years)	44.04	1.14	42.63	2.85	44.47	1.22		
Education level of household head (years)	2.90	0.15	3.03	0.30	2.86	0.18		
Household size (number)	7.75	0.31	7.40	0.52	7.86	0.38		
Number of men older than 16	1.86	0.11	2.23	0.22	1.75	0.12		
Number of women older than 16	1.74	0.10	1.71	0.17	1.75	0.12		
Number of people 16 or younger	4.15	0.23	3.46	0.33	4.36	0.28		
Experience with cotton (years)	14.68	1.04	16.86	2.36	14.02	1.15		
Yield loss from bollworm (percent)	0.74	0.35	0.59	0.38	0.78	0.33	2.8926	0.004
Land value (US$)	1,192.90	2,330.54	1,167.10	1,634.88	1,200.68	2,508.77		
Total area (hectares)	1.42	2.42	1.34	2.51	1.45	2.40		
Cotton area (hectares)	0.68	0.55	0.44	0.24	0.75	0.59	3.0129	0.003
Seed cotton price (US$ per kilogram)	0.39	0.06	0.39	0.08	0.39	0.05		
Seed cotton yield (kilograms per hectare)	953.48	719.66	675.53	548.62	1037.34	745.61	2.6592	0.009
Output value (US$ per hectare)	630.37	805.60	288.97	253.32	733.37	883.95	2.9320	0.004
Dummy for organic producer	0.34	0.48	0.34	0.48	—	—		
Dummy for use of herbicides	0.09	0.29	—	—	0.09	0.29		
Herbicide use (US$ per hectare)	1.38	5.70	—	—	1.38	5.70		
Dummy for use of fertilizer	0.09	0.29	0.03	0.17	0.11	0.32		
Fertilizer use (US$ per hectare)	1.01	7.06	0.04	0.23	1.30	8.04		
Dummy for use of pesticides	0.97	0.16	0.89	0.32	1.00	0.00	3.8431	0.000
Pesticide use (US$ per hectare)	21.52	20.89	9.42	26.23	25.16	17.55	4.1079	0.000
Labor for weeding (US$ per hectare)	69.41	70.89	66.79	86.11	70.20	66.03		
Labor for herbicide application (US$ per hectare)	0.18	0.83	—	—	0.18	0.83		
Labor for pesticide application (US$ per hectare)	6.94	10.86	4.31	7.91	7.73	11.51		
Total hired labor (US$ per hectare)	147.52	127.98	164.62	179.92	142.35	108.07		

Source: Authors' survey data.

Notes: t-test and P-value included only when significant. — = not applicable.

national average of about 400 kilograms per hectare reported by FAOSTAT for Uganda in 2007, productivity values are quite low.

Although household characteristics in Lira and Kasese are similar, some production variables are significantly different. Household size and composition, and the household head's age and level of education, are relatively similar across sites. Nor is there significant variation concerning land value, labor use, and years of experience in cotton cultivation. On average, farmers interviewed have more than 14 years of experience working with cotton. The size of cotton areas tends to be larger in Kasese. Similarly, seed cotton yield and output value generated from cotton production are also higher in Kasese. These results correspond to reports by local institutes. In contrast, it seems that the susceptibility to cotton bollworm is higher in Kasese (in Western Region) than in Lira (in Northern Region). According to cotton experts, this behavior can be extended to the regions, implying that Bt cotton could have greater success in Western than in Northern Region. Average productivity is also higher in Kasese. Interestingly, organic insecticides are used by the majority of producers both in Kasese and Lira. However, the average investment in pesticides is significantly different between the districts, being higher for Kasese.

If the Ugandan government decides to approve the introduction and commercialization of GM cotton varieties, a geographic segregation of areas planted to GM and non-GM cotton could be an alternative to take into consideration. Northern Region and other regions where organic cotton production is important could remain as GM-free areas.

To get some insights about farmers' behavior with respect to GM seed technology, we artificially classified producers as "low-input" and "high-input" users. Obviously this classification is also a proxy for income level. Because basically all conventional producers applied insecticides but very few applied mineral fertilizers, we used chemical fertilizer application as a criterion to classify producers. Therefore, the category "high-input user" refers to farmers that use chemical fertilizers and more than the average amount of pesticides. From a total of 151 observations, only 27 qualified as high-input users. All organic producers were categorized as low-input users. Table 5.4 presents the descriptive statistics of key variables by type of producer. We noticed that there were significant differences between low-input and high-input users in level of experience and total labor used but not in total area. In other words, high-input producers in our sample were using more inputs (chemical fertilizers, insecticides, and labor), independently of farm size.

Note also that the price of seed cotton obtained by high-input users was statistically higher than the one received by low-input users. Clearly, better quality produce fetched a higher price and allowed a higher investment in inputs. Accessibility is often the main reason for high-input use: farmers who are closer to markets tend to use more inputs when available. The total magnitude of investment in inputs was probably the result of either shorter distances to markets or other points of sale or better access to credit sources. Either way, an intervention is needed to guarantee farmer accessibility to good quality inputs and thus increase the use of productive inputs not only among low-input users but also among all types of producers. Low-input users could face similar constraints in reaching GM cotton technology as for any other productive inputs. Characteristically, credit can be particularly critical for cotton production because of the high use and cost of pesticides and the high labor demands throughout the planting season.

Gender of the household head and asset control can affect adoption and management of GM varieties (Table 5.5). Female-headed households probably have different priorities in the technology adoption process than

TABLE 5.4 Descriptive statistics for farm households by type of producer

Variable	Low-input users (N = 124)		High-input users (N = 27)		
	Statistic	Standard error	Statistic	Standard error	F
Gender of household head (female = 1)	0.08	0.27	0.15	0.36	
Age of household head (years)	43.49	13.39	46.22	17.33	
Education level of household head (years)	2.77	1.86	3.44	1.89	
Land value (US$ per hectare)	2,568.01	5,708.80	3,929.03	4,523.43	
Total area (hectares)	1.30	2.31	1.90	2.80	
Cotton area (hectares)	0.69	0.58	0.64	0.30	
Experience with cotton (years)	13.73	12.30	18.81	14.30	3.6*
Yield loss from bollworm (percent)	0.74	0.35	0.67	0.36	
Seed cotton price (US$ per kilogram)	0.38	0.04	0.41	0.03	31.3***
Output value (US$ per hectare)	583.81	808.54	713.27	580.04	
Seed cotton yield (kg/ha)	918.46	715.10	1,132.78	705.57	
Labor used for weeding (US$ per hectare)	60.57	65.07	95.70	75.28	3.6**
Total labor used (US$ per hectare)	132.45	112.67	188.25	157.35	3.0**

Source: Authors' survey data.
Notes: * indicates significance at the 10 percent level, ** indicates significance at the 5 percent level, and *** indicates significance at the 1 percent level.

do male-headed households. Female-headed households are usually not common and could easily be underrepresented in a sample. The household head, however, does not necessarily manage all the plots. In our case, although the percentage of female household heads was low (3 percent in Lira and 11 percent in Kasese), the share of plots managed by women was 51 percent in Kasese and 29 percent in Lira. Interestingly, when testing for mean differences between plots managed by men or women, none of the variables included in Table 5.5, except for gender and age of the household head, were significantly different.

In contrast to our finding, Baffes (2009) reports a large productivity gap for the southeastern cotton-producing region, with male growers often achieving yields three or four times higher than that of their female counterparts. We speculate that these differences are due more to the approach used to evaluate gender issues than to geographic differences. In other words,

TABLE 5.5   Descriptive statistics for farm households by gender

Variable	Male-controlled plot (N = 112)		Female-controlled plot (N = 24)		F
	Statistic	Standard error	Statistic	Standard error	
Gender of household head (female = 1)	0.03	0.02	0.46	0.10	55.56***
Age of household head (years)	42.88	1.36	50.13	2.70	5.16**
Education level of household head (years)	2.93	0.18	2.42	0.34	
Land value (US$ per hectare)	2,660.02	471.26	2,782.70	514.64	
Total area (hectares)	1.38	0.21	1.35	0.20	
Cotton area (hectares)	1.62	0.11	1.35	0.15	
Experience with cotton (years)	14.52	1.24	15.04	2.37	
Yield loss from bollworm (percent)	0.71	0.03	0.76	0.06	
Seed cotton price (US$ per kilogram)	0.38	0.00	0.38	0.00	
Output value (US$ per hectare)	535.44	51.97	554.97	58.71	
Seed cotton yield (kilograms per hectare)	932.60	63.28	949.38	71.54	
Labor used for weeding (US$ per hectare)	66.87	6.00	66.50	6.89	
Total labor used (US$ per hectare)	140.54	10.91	140.85	12.61	

Source: Authors' survey data.

Notes: ** indicates significance at the 5 percent level, and *** indicates significance at the 1 percent level.

although female-headed households are getting lower yields than male-headed households (Baffes 2009), these differences might be unobservable when the unit of comparison is not the head of household but the person managing the plot. Both comparisons have important implications for GM adoption and impact. Comparing household heads by analyzing the impact of GM technologies on disadvantaged female-headed households is essential to discussing poverty reduction and welfare effects (Meinzen-Dick et al. 2010). Comparing plot managers by assessing plot management strategies will help us understand differences in the decisionmaking process with respect to adoption of GM technologies.

Yield Determinants

According to the primary information collected, the main inputs used in cotton production in the selected sites are seeds, chemical and organic fertilizers, chemical and organic pesticides, herbicides, and labor. As discussed earlier, seed is distributed by ginneries, and farmers do not directly pay for it. Seed quality could be a constraint, depending on the previous year's harvest. Use of organic fertilizers and organic pesticides occurs mainly in Lira, where organic production is concentrated. Use of chemical fertilizers and pesticides is more common in Kasese. The use of these inputs is rather limited, however. Our survey shows that in 2007, only six farmers used chemical fertilizers and only three of them used an organic fertilizer. Table 5.6 shows the results of the production function regression using the damage-control specification. This table shows that fertilizer does not have a significant effect on yield performance (low t-values). Cotton is a labor-intensive crop, especially when produced organically, and labor, both family and hired, contributes significantly to yield performance.

In the damage-control part of the production function (Table 5.6), neither pesticides nor herbicides seem to be controlling damage caused by pest and weeds to any significant degree, as shown by the lack of statistical significance of these variables. This demonstrates the current underutilization of these inputs. Thus, investment in fertilizers, good quality seed, and other inputs is crucial to improving the poor cotton performance in Uganda. Although the introduction of GM seed could help control bollworm and weeds and reduce the amount of labor used for weeding, cotton would hardly achieve its full yield potential if the applications of fertilizer and other productive inputs remain at their current levels.

TABLE 5.6   Production function using a damage-control specification

Explanatory variable	Coefficient	Standard error	t-value
Production function			
Constant	198.56	37.72	5.26***
District (dummy, Kasese = 1)	219.46	83.22	2.64***
Altitude (miles above sea level)	−0.25	0.09	−2.76***
Organic producer (dummy)	242.5	171.51	1.41
Land rent (US$ per hectare)	−0.61	1.37	−0.45
Square of land rent	0	0	0.73
Family labor (US$ per hectare)	0.38	0.16	2.41**
Square of family labor	0	0	−2.07**
Fertilizer (US$ per hectare)	11.61	9.56	1.21
Square of fertilizer	0	0	−0.30
Hired labor for harvesting (US$ per hectare)	2.19	1.66	1.32
Square of hired labor for harvesting	0	0	−0.16
Hired labor for other activities (US$ per hectare)	4.72	0.83	5.68***
Square of hired labor for other activities	0	0	−4.32***
Damage control			
Constant	10.54	10.57	1.00
Pesticide and labor used to apply pesticides (US$ per hectare)	0.57	0.57	1.01
Herbicide and labor used in weeding (US$ per hectare)	0.07	0.07	0.95

Source: Authors' survey data.

Notes: ** indicates significance at the 5 percent level, and *** indicates significance at the 1 percent level. $R^2 = 0.45$; adjusted $R^2 = 0.38$.

Production Costs

The production costs for both conventional and organic production systems are presented in Table 5.7. Even though the organic system has a slightly higher benefit-cost ratio, both systems have low profitability, especially after including family labor. Cotton production is generally a risky business: note that the downside risk (that is, the probability of having a negative profit) is high for both systems.

Conventional System

Cotton is a low-profitability activity in Uganda regardless of the production system (Table 5.7). Although the reported productivity of seed cotton for the sample (about 900 kilograms per hectare) is above the reported national average (about 400 kilograms per hectare), the benefit-cost ratios estimated are

TABLE 5.7   Cotton profitability, 2007/08 season

Income or cost component	Conventional (N = 139)	Share (percent)	Organic (N = 12)	Share (percent)
Yield (kilograms per hectare)	962.05	—	863.46	—
Price reported by farmers (US$ per kilogram)	0.38	—	0.38	—
Total income (US$ per hectare)	366.76	—	328.21	—
Land rent (US$ per hectare)	74.99	22	72.25	26
Chemical fertilizer (US$ per hectare)	24.56	7	0.00	—
Organic fertilizer (US$ per hectare)	20.23	6	22.01	8
Herbicide use (US$ per hectare)	17.16	5	0.00	—
Pesticide to control lepidoptera (US$ per hectare)	23.61	7	—	—
Pesticide to control other insects (US$ per hectare)	17.44	5	—	—
Organic pesticide (US$ per hectare)	—	—	4.82	2
Hired labor to apply pesticides (US$ per hectare)	7.24	2	3.37	1
Hired labor to apply herbicides (US$ per hectare)	5.42	2	0.00	—
Hired labor for weeding (US$ per hectare)	70.91	20	61.20	22
Hired labor for harvesting (US$ per hectare)	28.92	8	28.90	10
Hired labor for other activities (US$ per hectare)	57.49	17	89.09	32
Family labor (US$)	423.98	—	1,652.39	—
Total costs (US$ per hectare)	347.98	—	281.66	—
Margin (US$ per hectare)	18.78	—	46.56	—
Downside risk (percent)	38.60	—	52.00	—
Benefit-cost ratio	1.05	—	1.17	—

Source: Authors' survey data.

Notes: Values presented are the mean values of field observations. An average wage rate of US$2.90 per day is used to esti-mate the opportunity costs of family labor. When family labor is added to the estimations in the table, the benefit-cost ratio decreases significantly, falling to 0.48 for conventional producers and to 0.17 for organic producers. — = not applicable.

still low.[5] Conventional producers do get a higher benefit-cost ratio, but the returns are very low when family labor is taken into account. Moreover, the opportunity costs of family labor could be underestimated, as family labor estimates are drawn from farmers' recollections and their subjective valuation (use) of their own labor.

Under conventional production, most farmers use some type of chemical control to deal with insect pests. Relatively few use fertilizers, and almost none use herbicides. This last input could contribute significantly to improving crop

---

5 According to FAO (2010), average productivity for seed cotton has been about 417 kilograms per hectare for the past 5 years, and for the past 10 years it has been about 347 kilograms per hectare.

profitability. Cotton is a labor-intensive crop; labor represents about 50 percent of total production costs—more than 50 percent in the case of organic producers. Most labor is used for manual weeding. Weed infestation is therefore another severe constraint on cotton production. In the study sample, weeding represents about 20 percent of the total labor costs for both types of producers. Other institutions working in cotton in the area report similar patterns (Agricultural Productivity Enhancement Program, personal communication, 2008).[6]

In short, cotton is a labor-consuming activity, and weeding is particularly labor demanding. So, freeing labor from weeding could allow family members to be available for other economic activities. In contrast, freeing labor through the use of HT seed could have a negative impact on employment and welfare in the community if there are no productive off-farm labor opportunities.

Farmers in Kasese and Lira seem to have serious problems with bollworm. On average, this pest can damage up to 70 percent of expected output. Although these estimations are based on farmers' perceptions and may have an upward bias, they are a good reference point for understanding the severity of bollworm infestation in these regions. In addition to bollworm, there are other common biotic stresses, such as aphids, *Lygus* spp. (sucking-type insects), and cotton strainers. The severity of these biotic constraints, combined with high price variability and poor access to inputs, has transformed cotton production into a highly risky activity in Uganda. The estimated downside risk (nearly 40 percent) for surveyed conventional farmers illustrates the magnitude of the problem.

Organic System

One of the purposes of the implementation of household surveys in Lira was to cover organic producers and collect information to generate a representative partial budget for an average organic cotton producer. Nevertheless, only 12 of the 35 self-identified organic producers qualified as actual standard complying organic producers. The remaining 23 admitted using some chemicals to control heavy pest infestations. In fact, the number of organic farmers appears to change from year to year, as farmers seem to switch from conventional to organic with relative ease. According to the company Dunavant, in the 2006/07 season, 11,691 organic farmers were registered and contracted for a total production of 6,600 bales (about 185 kilograms per bale),

---

6 Survey results do not support the general notion that women manage cotton agronomic practices (weeding in particular).

which accounted for almost one-third of Dunavant production. During the 2008/09 season, there were serious problems with the production of organic cotton, because army bollworms infested the crop.

Even though it is not possible to make statistical inferences based on a small number of observations, the analysis of household survey information can provide some useful insights. It is well known, for instance, that the profitability of organic cotton is significantly low (Ogwang, Sekamatte, and Tindyebwa 2005). For the sampled farmers, the marginal benefits are less than 17 percent of total costs. In addition, the downside risk—the risk of not being able to cover at least the production costs—is higher than 50 percent (see Table 5.7).

As with conventional production, organic cotton faces several biotic and abiotic constraints. Surveyed farmers report that the damage caused by bollworm is greater than 50 percent (see Table 5.3). Similar to conventional cotton, the main cost in organic cotton production is labor (65 percent). Organic production generally requires a significant amount of labor for manual activities, including insect and weed control. Our study does not have information about specific control practices implemented.

Simulation Results

The profitability of cotton production is low for all the simulations and scenarios accounting for the two different technology fees. Also, none of the simulations show first-degree stochastic dominance over the others, which confirms that none has an outcome that is clearly better on average than the others. Note also that the downside risk of each alternative is high, but it is particularly high for the scenario with a higher technology fee.

Bt and HT cotton are good alternatives that may partly reduce the risk in cotton production, especially when the technology fee is low (Table 5.8, technology fee 1). However, the reduction of downside risk depends on the effectiveness of the GM technology in controlling the constraint (expression of the trait). Experts report that yield losses stemming from bollworm could be as high as 80 percent, which is in agreement with what farmers have reported (on average, about 76 percent). Given these high values attached to farmers' perceptions about yield losses from bollworm attack and weed infestation, it is not surprising that the margins are higher for both the Bt cotton and HT cotton scenarios than they are for the organic cotton scenario with technology fee 1. But perceptions are usually upward biased, given that it is rather difficult for farmers to isolate the net effect of one constraint from all the other constraints they face.

The results of the simulation show that the marginal benefits and the benefit-cost ratio are higher when GM seed is used. The marginal benefits of using GM seed are directly related to the level of incidence of the productivity constraint and the actual damage caused by the biotic constraint. This is particularly interesting, because the average technology efficacy is 50 percent, meaning that the GM seed is able to control at least 50 percent of the yield losses caused by bollworm and by weeds. To generate the distribution for technology efficiency, we allowed the efficacy rate to vary from no efficiency at all (0 percent) to full control of the constraint (100 percent efficiency). The literature shows that this is a sensible assumption (Qaim and Zilberman 2003).

For the HT cotton simulation, the assumptions are based on expectations. This simulation records the highest benefit-cost ratio and the highest marginal rate of return over low-input or organic production systems. Unfortunately, it also produces the least solid results because of lack of technical information (for example, the number of weeding sessions avoided with one application of the herbicide Roundup® [Monsanto, Marysville, OH]), yield loss due to weeds, and the comparatively low number of respondents. Unlike Bt cotton, which has been documented relatively well, there is only scattered technical information about the performance of HT cotton in farmers' fields.

Organic cotton is relatively attractive for farmers who already use few inputs and no chemical pesticides and can gain from receiving a price premium. According to public sources, this premium can be as high as 20 percent more than the usual price per bale of seed cotton (ACE 2006), although prices reported by farmers in Lira do not seem to be considerably higher than those received by conventional producers. This difference could be explained by transportation and ginnery fees charged to producers, because farmers report total income from selling their cotton production rather than price per bale of seed cotton. Table 5.8, fifth column, illustrates a simulation of organic production with an effective price premium that results in better cotton profitability (benefit-cost ratio) than the one reported for organic producers in Table 5.7. Nevertheless, given the high downside risk of organic cotton and its low profitability, it is a far less appealing option for farmers than producing GM cotton with a low technology fee. However, when the technology fee of GM cotton is at international levels (technology fee 2 in Table 5.8), the profitability of organic cotton is higher than in any other simulation, including the simulation where Bt cotton seed is used in an organic system. The latter actually recorded the highest downside risk (55.6 percent), which is explained partly by the high yield variability of organic cotton and partly by the limited number of observations for organic producers.

TABLE 5.8   Partial budget simulations

Cost component	Bt cotton	HT cotton	Organic + premium price	Organic + Bt cotton
Yield (kilograms per hectare)	1,325.40	1,342.60	863.50	1,101.34
Yield loss from bollworm (percent)	76.00	—	55.00	55.00
Yield loss from weeds (percent)	—	79.00	—	—
Technology efficacy (percent)	50.00	50.00	—	50.00
Price reported by farmers (US$ per kilogram)	0.38	0.38	0.43	0.43
Premium price (US$ per kilogram)	—	—	12.50	12.50
Total income (US$ per hectare)	505.27	511.83	369.24	470.97
Seed cost[a] (US$ per hectare)				
Technology fee 1	1.58	1.58	0.00	1.58
Technology fee 2	128.00	128.00	0.00	128.00
Land rent (US$ per hectare)	74.99	74.99	72.25	72.25
Chemical fertilizer (US$ per hectare)	24.56	24.56	0.00	0.00
Organic fertilizer (US$ per hectare)	20.23	20.23	22.01	22.01
Herbicide use (US$ per hectare)	17.16	25.74	0.00	0.00
Rate of herbicide use increase (percent)	—	50.00	—	—
Pesticide to control lepidoptera (US$ per hectare)	11.81	23.61	—	—
Rate of pesticide use reduction (percent)	50.00	—	—	—
Pesticide to control other pests (US$ per hectare)	17.44	17.44	—	—
Chemical pesticide (US$ per hectare)	—	—	0.00	0.00
Organic pesticide (US$ per hectare)	—	—	4.82	4.82
Labor to apply pesticides (US$ per hectare)	5.43	7.24	3.37	2.53
Rate of labor cost reduction (percent)	25.00	—	—	25.00
Labor to apply herbicides (US$ per hectare)	5.42	8.13	0.00	0.00
Rate of increase in labor costs (percent)	—	50.00	—	—
Labor for weeding (US$ per hectare)	70.91	35.45	61.20	61.20
Rate of labor cost reduction (percent)	—	50.00	—	—
Labor for harvesting (US$ per hectare)	28.92	28.92	28.90	28.90
Labor for other activities (US$ per hectare)	57.49	57.49	89.09	89.09
Family labor (US$ per hectare)	423.98	423.98	1,652.39	826.19
Technology fee 1				
Total costs (US$ per hectare)	335.94	325.40	281.66	282.39
Margin (US$ per hectare)	169.33	186.44	87.58	188.57
Downside risk (percent)	26.80	21.60	48.40	40.80
B-C ratio	1.50	1.57	1.31	1.67
Technology fee 2				
Total costs (US$ per hectare)	462.36	451.82	Same as above	408.81
Margin (US$ per hectare)	42.91	60.02		62.15
Downside risk (percent)	43.90	38.70		55.60
B-C ratio	1.09	1.13		1.15

Source: Authors' survey data.

Notes: Numbers are rounded estimates. When family labor is added to the estimations using the average wage rate (US$2.90 per day), some changes occur. In the case of technology fee 1, the B-C ratio decreases to 0.66 for Bt cotton, 0.68 for HT cotton, 0.19 for organic production that can get at least half of the premium price offered, and 0.42 for a production that combines organic management with Bt cotton technology. In the case of technology fee 2, the B-C ratio decreases to 0.57 for Bt cotton, 0.58 for HT cotton, and 0.38 for production that combines organic management with Bt cotton technology. B-C ratio = benefit-cost ratio; Bt = insect resistant; HT = herbicide tolerant; — = not applicable.

[a] Seed is assumed to be applied at the rate of 10 kilograms per hectare.

FIGURE 5.1 Distribution and sensitivity of marginal benefits for cotton producers

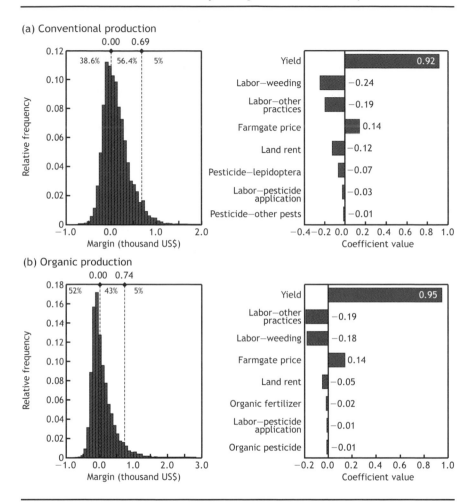

(a) Conventional production

(b) Organic production

In terms of productivity, conventional producers are expected to perform better than organic producers, as shown in Table 5.8. Among probable production options, the rate of returns is higher for farmers who make use of chemical inputs and pay a low technology fee, even considering a premium price for organic cotton. However, in reality, organic producers do not seem to be getting a premium price for their product. If there is no premium price, then there are no marginal returns that will provide incentives to farmers to move from low-input production to organic production. If there is a premium

FIGURE 5.1   (continued)

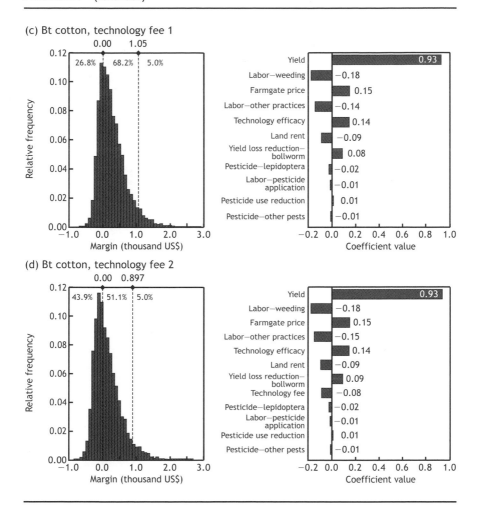

(c) Bt cotton, technology fee 1

(d) Bt cotton, technology fee 2

Source: Authors' survey data.

price and if the technology fee for GM seed is set at international standards and fully charged to farmers, then the marginal returns to organic production are the highest across all simulations.

Figure 5.1 shows both histograms and tornado graphs for eight production alternatives. The histograms present the distribution of marginal benefits of real and simulated production alternatives: conventional, organic, conventional using Bt cottonseed, conventional using HT cottonseed, and organic using Bt cottonseed. See Appendix 7 to find the graphs for the

FIGURE 5.1   (continued)

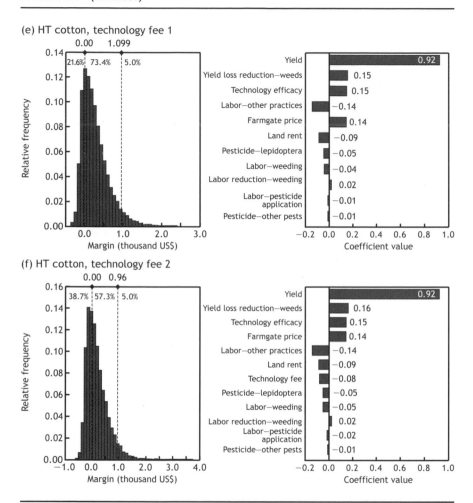

(e) HT cotton, technology fee 1

(f) HT cotton, technology fee 2

Source: Authors' survey data.

organic plus premium price scenario. The tornado graphs present the sensitivity of marginal benefits to the different input variables that use a probability distribution in the simulation. The @Risk software generates these tornado graphs by estimating a simple linear regression using marginal benefits (or any other output variable) as the dependent variable and all the input variables substituted with probability distributions as explanatory variables. The value of each coefficient can be interpreted as a measure of how much the output (marginal benefits) would change if the input variables

FIGURE 5.1   (continued)

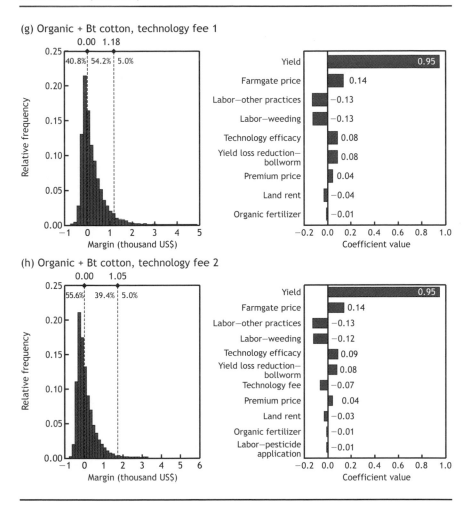

(g) Organic + Bt cotton, technology fee 1

(h) Organic + Bt cotton, technology fee 2

Source: Authors' survey data.

(such as yield or technology fee) were changed by the equivalent of one standard deviation.

Note that a subsidized technology fee (technology fee 1) does not have a significant effect on marginal benefits. This situation changes when the technology fee is set at international levels (technology fee 2). Across all production alternatives, the variability in yield and the high labor costs are the main determinants of the margins generated. A technology that contributes to reducing yield variability would have a definitive impact on farmers' welfare.

Sensitivity Analysis

The adoption of an agricultural technology depends critically on the degree to which the technology is able to reduce risk and uncertainty in production.[7] Two of the greatest sources of uncertainty and risk in rainfed agriculture in Africa south of the Sahara are future rainfall patterns and output prices, which can have consequences involving income loss, missed opportunities for increasing income, or both (Kelly 2006). The sensitivity analysis presented here deals with risk but not with uncertainty: we use change in marginal benefits as a risk indicator. Reducing risks can change farmers' perceptions of a technology and increase the incentives to adopt it; thus, farmers can better capture the full potential of the technology.

The results of our simulations in Table 5.8 are subject to a number of assumptions that allowed us, as much as possible, to insert variability into the simulations. The assumptions are based either on expert information or on information from highly regarded publications. However, some aspects of the biology and agronomic performance of the crop are harder to simplify into general assumptions, because they involve complex relationships and interactions. Under ex ante conditions, where there is no technical information available, the situation is somewhat more complicated. Accounting for the effect of inputs' complementarity on GM cotton performance is a clear example of this situation. It is important, however, to at least have a picture of the potential interaction among key production inputs (such as labor, mineral fertilizers, and chemical insecticides) and in this context to understand the role and the need for a reliable and affordable source of credit.

WHAT WOULD BE THE EFFECT OF GM ADOPTION ON LABOR DEMAND?

Cotton is a labor-intensive crop. Farmers in Uganda use family labor and quite frequently use hired labor too. As shown in the analysis, family labor is not accounted for in our budget estimations. The use of GM cotton varieties would probably demand higher labor for agricultural practices, mainly for harvesting. Initially, this additional labor for harvesting could be done by family members. This is a strong possibility for low-input users, who make up the majority both in our sample and in Uganda. Another strong possibility is partial adoption, where low-input users would plant GM cotton in part of the plot. This partial adoption may not require additional hired labor.

---

7   Hardaker et al. (2004) refer to uncertainty as "imperfect knowledge" and to risk as "uncertain consequences."

In general, many factors can affect labor demand in GM cultivation. For budget estimations, it can reasonably be assumed that a higher yield will produce a proportional increase in labor demand, although doing so can result in overestimating, because either labor may not be readily available or the household may lack enough financial resources to hire additional labor. In any case, as shown in Table 5.8, yield increases imputed to GM seed are basically due to the efficiency of the technology in abating damage (the damage being the expected yield loss due to bollworm multiplied by GM technology efficiency). If yield shows a positive and perfect correlation to labor use for harvesting, then the relative yield increase can be used as an upper bound on additional labor requirements for harvesting.

Using the average values in Table 5.8 to recalculate partial budgets in Table 5.9, we notice that when potential changes in labor demand are taken into account, the reduction in marginal benefits is more severe under the high technology fee (technology fee 2 in the table). Bt cotton reports the highest reduction in marginal benefits under either technology fee scenario, whereas the organic + Bt scenario reports the lowest reduction. In other words, it is important to take into account the effect of higher labor demand on expected profits. If the producer uses mainly hired labor, it will definitely have an impact on profit margins. It is expected that this type of producer will also invest in other complementary inputs that will contribute to an even better performance of the variety and thus compensate for the hired labor costs and additional investment in other inputs.

WOULD THE USE OF MINERAL FERTILIZERS AUGMENT THE PROFITABILITY OF GM COTTON?

A major factor in cotton production is the availability of adequate and balanced nutrients. Fertilizers are underused in cotton in Uganda, despite the common belief that chemical fertilizers are primarily used for export crops. In general, Yanggen et al. (1998) explained that cotton in Africa south of the Sahara has relatively poor yield response to fertilizers and mediocre profitability. They attribute this behavior to a lack of adequate incentives in terms of yield, price, or expected profit (Kelly 2006). In other words, farmers do not perceive the advantage of using fertilizers, because their use (or increased use) does not necessarily translate into higher yields, better prices, or better profits.

The effect of fertilizers is very complex, because it depends on a number of environmental and management practices. Better cotton performance is the expected result of using chemical fertilizer. However, we do not have enough technical information to make assumptions about soil fertility and its ability to

TABLE 5.9   Sensitivity analysis for complementary inputs

Scenario	Bt cotton Margin (US$ per hectare)	Change (%)	B-C ratio	HT cotton Margin (US$ per hectare)	Change (%)	B-C ratio	Organic + Bt cotton Margin (US$ per hectare)	Change (%)	B-C ratio
Baseline									
Technology fee 1	169.33	—	1.50	186.44	—	1.57	188.57	—	1.67
Technology fee 2	42.91	—	1.09	60.02	—	1.13	62.15	—	1.15
Labor for harvesting									
Technology fee 1	158.41	−6.45	1.46	175.00	−6.14	1.52	180.61	−4.22	1.62
Technology fee 2	30.36	−29.25	1.06	48.58	−19.06	1.10	54.19	−12.81	1.13
Fertilizer[a]									
Yield increase of 10%									
Technology fee 1	195.60	15.51	1.54	212.67	14.07	1.61	—	—	—
Technology fee 2	67.39	57.03	1.14	86.25	43.71	1.18	—	—	—
Yield increase of 25%									
Technology fee 1	251.38	48.45	1.66	269.18	44.38	1.73	—	—	—
Technology fee 2	122.93	186.45	1.24	142.76	137.86	1.29	—	—	—
Yield increase of 50%									
Technology fee 1	344.35	103.36	1.83	363.36	94.89	1.90	—	—	—
Technology fee 2	215.49	402.15	1.40	236.93	294.77	1.45	—	—	—
Yield increase of 75%									
Technology fee 1	437.33	158.26	1.98	457.53	145.41	2.04	—	—	—
Technology fee 2	308.06	617.85	1.54	331.11	451.69	1.59	—	—	—
Yield increase of 100%									
Technology fee 1	530.30	213.17	2.10	551.71	195.92	2.17	—	—	—
Technology fee 2	400.62	833.55	1.66	425.29	608.61	1.71	—	—	—
Pesticide[b]									
Technology fee 1 + fertilizer									
Yield increase of 10%	184.08	8.71	1.50	208.31	11.73	1.59	—		
Yield increase of 25%	239.86	41.65	1.61	264.82	42.04	1.71	—	—	—
Technology fee 2 + fertilizer									
Yield increase of 50%	197.52	360.27	1.35	232.57	287.51	1.43	—	—	—
Yield increase of 75%	290.09	575.79	1.49	326.75	444.43	1.57	—	—	—
Yield increase of 100%	382.65	791.67	1.61	420.93	601.35	1.70	—	—	—

Source: Authors' survey data.

Notes: Numbers are rounded estimates. B-C ratio = benefit-cost ratio; Bt = insect resistant; — = not applicable.

[a]These estimations include the costs associated with the additional labor for harvesting.

[b]Fertilizer and labor changes are included. Pesticides control the damage of secondary pests, so the effect of fertilizer is realized.

use the minerals provided by fertilizer. This type of information is location sensitive and, given the high soil diversity and heterogeneous conditions in Africa south of the Sahara, it is difficult to make generalizations. Even the standard fertilizer recommendations are not appropriate for this region, and more site-specific recommendations for increasing productivity are needed (Carr 1993).

The response of the GM crop varieties to fertilization is unknown, and the variability across producers will probably be high. Instead of guessing at the magnitude of this effect, we set targets for yield increases (10, 25, 50, 75, and 100 percent) resulting from fertilizer application. We use the output-nutrient ratio reported in Kelly (2006) to estimate marginal budgets for different yield responses to fertilization. A yield response is the additional yield obtained with one additional unit of fertilizer; yield increases are a percentage change over the baseline. Kelly (2006) reports typical, minimum, and maximum values for a yield response. We use the typical or modal value (5.8) mainly to avoid the influence of outliers.

Table 5.9 presents marginal benefits under different levels of fertilization. These results show that marginal benefits and the benefit-cost ratio improve relative to the baseline for GM cotton even when the target is only a 10 percent yield increase from fertilizer application, and especially when the technology fee is subsidized (technology fee 1). When the technology fee is fully paid by farmers (technology fee 2), investment in fertilizer has to be much higher compared to scenario 1 to obtain comparable benefits. These findings indicate that fertilizer use in cotton production is low partly because of the large (and expensive) amount required to make production profitable. The results highlight the importance of negotiating a fair technology fee and making credit available to farmers so they can pay it.

WOULD GM SEED REDUCE PESTICIDE USE AND INCREASE THE MARGINAL BENEFITS OF COTTON?

The relationship between GM cotton and pesticide use is complicated, especially in the case of Bt cotton. GM cotton is just as vulnerable as any non-GM variety to drought or outbreaks of non-bollworm pests. In China, India, and South Africa, reductions in pesticide use, related mainly to the use of Bt cotton, have been observed. However, there is some evidence that initial reductions in pesticide use have been reversed when insecticides are needed to control secondary non-bollworm pests (Pemsl, Waibel, and Orphal 2004; Glover 2010). As a result, some farmers in China and India who have adopted GM cotton varieties continue to apply excessive quantities of insecticides (Pemsl, Waibel, and Orphal 2004). As in China and India, farmers in Africa have to deal with pesticides of uncertain origin and quality (Pemsl, Waibel, and Gutierrez 2005; Tripp 2009). Thus, as argued by Glover (2010), farmers face a difficult and uncertain set of trade-offs when determining their risk management strategies. As a consequence, these farmers opt to keep using high levels of insecticides. This can happen in Uganda as well.

In Uganda, in contrast with the situation in China or India, our analysis of the production factors shows that neither insecticides nor herbicides are used in optimal quantities (see Table 5.6, damage-control analysis). Under these circumstances, the adoption of Bt cotton is more likely to be accompanied by an increase, rather than a reduction, in the use of insecticides to control secondary pests. To depict this situation, we estimated a partial budget that assumes a 25 percent increase in the use of insecticides to control other nonlepidopteran pests. Because additional insecticide applications will require more labor, we discarded the 25 percent labor-reduction effect originally considered in the baseline estimation (see Table 5.8). The partial budget estimation for Bt cotton also assumed a reduction in insecticide use against lepidoptera.

In this last estimation, insecticides are treated as damage-control inputs, and for this reason we do not assume a yield increase stemming from insecticide use. Insecticides merely control secondary pests so the fertilizer's effect is realized. Marginal benefits are higher than the baseline but lower than just accounting for fertilizer's yield effect. The technology fee is again a key factor that determines not only the magnitude of benefits but also the level of investment in additional inputs needed to make GM cotton production profitable. For most cotton farmers, there are just not enough incentives to invest in high-quality inputs.

WHAT WOULD BE THE ROLE OF RURAL CREDIT?

Given this analysis of complementary inputs, what would be the role of rural credit? The results above only confirm the well-known fact that cotton producers will largely benefit from having access to credit. Kabwe and Tschirley (2007) provide evidence for the validity of this argument in Africa south of the Sahara, a region where (1) cotton production requires substantial use of production inputs, (2) smallholder farmers are typically cash constrained, (3) input markets are weak, and (4) rural credit markets for agriculture are nearly nonexistent.

Poor access to credit may be the main reason farmers are not adopting fertilizer and might not adopt GM seed. The analysis of input-use patterns done by Reardon et al. (1997) concluded that in Burkina Faso, Senegal, and Zimbabwe, the elimination of fertilizer credit and subsidies associated with structural adjustment programs led to sharp reductions in fertilizer use.

Another example is the case of South Africa, where the initial success of Bt cotton in the Makhathini Flats depended heavily on the joint support of the local cotton company and the local credit agency (Gouse et al. 2005; Glover 2010). Baffes (2009), however, warns about the possibility that in Uganda, the decision to grow cotton can reflect the high costs of credit only for some farmers, whereas for other farmers it reflects the absence of alternative sources of cash income. Baffes also concludes that policies based on the assumption that provision of credit is the key constraint are likely to lead to inadequate policy recommendations.

In our sample, about 28 percent of the farmers had access to some source of cash credit, and 55 percent of those with access had to pay interest on the loan, which ranged from 10 to 100 percent. Higher interests were usually linked to smaller loans, which created a clear disadvantage for smallholder and cash-constrained farmers. About 13 percent of the farmers in our sample had access to in-kind loans, receiving seeds or other production inputs to be repaid at harvest. In the history of Ugandan cotton production, there have been several efforts to support farmers with inputs and credit (Baffes 2009), the most recent one being the zonification abandoned in 2006. Given the current situation in Uganda, introducing a technology such as GM crops should be accompanied by setting up rural credit markets or alternative forms of farmer-organized credit that are accessible and affordable to farmers, particularly small producers. When baseline levels of uncontrolled pest damage are high, as in Uganda, GM cotton (namely, Bt cotton) can confer an unconditional yield effect in addition to the stochastic yield effect that is subject to environmental conditions (Lybbert and Bell 2010). This reduction in yield uncertainty could generate conditions that would encourage people, even small producers, to borrow, as well as to lend.

However, lack of credit is not the main constraint for all Ugandan cotton producers. There are areas, particularly in the Northern Region where organic cotton is produced, that are constrained mainly by the limited alternatives to cotton production as sources of cash income. In these areas, farmers' best option is probably to capture the higher prices fetched by organic cotton, although they could still derive benefits from better credit accessibility.

Conclusions and Policy Recommendations

This chapter has addressed the critical question of whether the adoption of GM cotton would make farmers in Uganda better off. A survey was used to collect farm-level information and to compare the profitability of various real and simulated production alternatives. The study evaluated six production alternatives: (1) conventional system, (2) organic system with current farm-gate prices, (3) organic system with a guaranteed premium price, (4) use of Bt cotton seed, (5) use of HT cotton seed, and (6) a hypothetical situation using Bt cotton seed in an organic system. The study also evaluated two scenarios for each of the production alternatives that involved the use of GM seed. In the first scenario, the technology is subsidized: that is, farmers do not pay the full price of the GM seed. In the second scenario, farmers assume the full costs of the technology.

At first glance, the estimated values of cotton profitability in this analysis do not seem to justify investment in a complex technology such as GM cotton. Overall, the simulations show that, with a subsidized technology fee (technology fee 1 in the tables), farmers using Bt and HT cotton varieties will achieve the highest returns, but the profitability of the crop will not increase dramatically compared to conventional or organic systems. If the technology fee has to be fully paid by adopting farmers (technology fee 2), then the adopters will most likely be wealthy farmers who would probably need to increase their investments in complementary inputs. In general, for all types of producers—particularly farmers adopting GM seed—appropriate access to complementary inputs, specifically fertilizer, will help improve the profitability (as measured by the marginal benefits) of cotton production.

Therefore, the government has two clear tasks here. The first one is related to the technology fee and to the need to guarantee that this fee is affordable for cotton producers. This task is crucial, especially if the goal of the technology introduction is to address poverty. The second task is to generate incentives for the provision of rural credit. Independent of the producer type, there is a need for access to an affordable source of credit and thus for access to complementary inputs. Evidence shows that fertilizer and pesticides do not significantly contribute to improved cotton performance. Although most producers interviewed have used some sort of chemical pesticide, few of them used fertilizer, despite its potential. The problem is not availability but mainly affordability. Cotton producers are often cash constrained and need access to some sort of credit to finance the crop campaign. For these producers, profits from cotton production simply do not justify the investment in costly inputs for a highly risky crop. GM cotton contributes to reducing the risks of crop production and thus to increasing the incentives for investment, but credit institutions need to provide access to affordable credit for GM cotton production to achieve the desired production and poverty reduction goals.

In the case of Bt and HT cotton, it is important that farmers are not using significant levels of insecticides and herbicides, and therefore the expected reduction of the use of both types of chemical would be insignificant. It is also important to take into account that if yield losses due to bollworm are lower than those self-reported by farmers (a situation known as "low incidence of the constraints" in the particular year), the profitability of this technology will dramatically decrease. For this reason, it is advisable to introduce Bt/HT cotton technology as a form of insurance, rather than as a way to increase yields. The technology will protect Ugandan farmers against severe or even catastrophic losses, because it targets pests and weeds.

As mentioned in Chapter 3, organic cotton production in Uganda, although unstable, is a growing industry. For this reason, it is important to take into consideration the potential impact of GM cotton adoption on organic production. Therefore, another goal of our study was to obtain a minimum sample of observations that would allow us to draw valid conclusions for the organic cotton sector. This proved to be more challenging than expected, as several of the identified organic producers reported using some amount of chemicals to control biotic damage.

It is possible to compare the profitability of a given year of organic cotton production with that of conventional cotton production using GM seed, but this provides only a partial view of the cotton landscape. Such information is valid for a rapid assessment and can help in decisionmaking, especially in the framework of a regulatory process and the time and budget constraints that this process implies. It is much more significant, but at the same time more challenging, to evaluate the long-term contribution of either an organic or conventional production system to farmers' welfare. This research topic requires further attention.

Chapter 6

Economic Impact on the Cotton Sector

José Falck-Zepeda, Daniela Horna, and Miriam Kyotalimye

This chapter assesses the potential aggregate net benefits for the cotton sector from the adoption of genetically modified (GM) cotton in Uganda, including both insect protection and herbicide tolerance in cotton.[1] We examine the magnitude and distribution of benefits, if any, among consumers, producers, and technology developers derived from the adoption of GM cotton. The method most often used to estimate the economic benefits of GM crops and other agricultural technologies is based on the economic surplus methodology outlined by Alston, Norton, and Pardey (1995). Other quantitative and qualitative methods have certainly been used for the ex ante assessment of agricultural biotechnologies: these other methods include linear programming, general and partial equilibrium models, and the real-options model.

In the case of GM cotton that expresses insect protection or herbicide tolerance, there is a need to carefully consider its damage-abatement nature, which is easily incorporated into an economic surplus approach. A real-options model can add irreversibility and uncertainty into the ex ante estimation of net benefits from the adoption of GM technologies, but the incorporation of damage abatement is more difficult.[2] In the same way, general equilibrium models can also be used to examine the potential adoption impact of GM crops, but these models require extensive modifications—and thus additional human and financial resources—to allow the possibility of shocking the computable general equilibrium (CGE) model by incorporating a specific agricultural technology's adoption.

Although the issues of irreversibility, uncertainty, and economywide and inter- and intramarket impacts are quite important in the assessment of agricultural biotechnologies, our study focuses on specific issues

---

1 Preliminary indications from experts consulted suggest that the GM cotton planned for introduction in Uganda would be insect-resistant (Bt) cotton.

2 A paper by Ndeffo Mbah et al. (2010) develops a model that incorporates the real-options approach to examine integrated pest-management strategies. We are not aware of a real-options model that addresses the issue of a Bt crop's damage-abatement characteristics.

relevant to insect resistance and herbicide tolerance, including the damage-abatement nature and the downside risk assessment. Moreover, the study considers the budget and time limitations that developing countries' practitioners may face in assessing GM crops within a biosafety regulatory process or a technology approval process. The economic surplus model, although quite limited in many aspects, is indeed parsimonious and flexible enough to allow rapid estimation of the impact of GM crop adoption in developing countries.

This chapter evaluates the potential benefits and risks derived from the adoption of cotton varieties that incorporate insect protection. In addition, we evaluate cotton varieties that incorporate insect protection and herbicide tolerance (also known as "stacked" varieties) in the Ugandan cotton market. We use the economic surplus model approach, enhanced through the use of stochastic simulation modeling techniques. In place of the static model parameters commonly used in the standard economic surplus model, the stochastic simulation procedure used in our study introduces probability functions that enable the assessment of both production and financial risk and addresses model parameter uncertainty. Stochastic simulations are used to account for the variability in the economic benefits stemming from changes in size and dispersion of the model parameters and thus to conduct a robust risk assessment. Furthermore, the chapter evaluates results estimated through the economic surplus model by using the stochastic dominance method.

Review of the Economic Surplus Model

The basic economic surplus model, as described by Alston, Norton, and Pardey (1995), consists of a set of supply and demand equations that model the market as a system. Algebraic manipulations of these equations allow computation of total surplus and its distribution into producer, consumer, and innovator surplus. In practice, the economic surplus model is implemented through the incorporation of equation parameters that include size and openness of the economy, size of the cotton sector, demand and supply elasticities, supply curve shifts, adoption rates, and years required for development and adoption. The basic economic surplus model may be modified to consider type of economy and trade, technology and input use biases, and government policies.

The reliability of economic surplus models depends largely on the extent to which the underlying parameters represent local conditions. In many ex ante evaluations, information regarding the different parameters is not readily available. To address this shortcoming, assessors conducting ex ante studies customarily use parameter values that come from the literature on related research conducted in other countries or are elicited from local experts. Because so few GM crop trait combinations have been released to date, there is little information available to researchers on which to base parameter identification and quantification. Parameter uncertainty limits the assessment of the potential impacts of crop biotechnologies in developing economies. We therefore use probability functions to enhance economic surplus models and thus to address parameter uncertainty.

The conventional economic surplus approach fails to account for the inherently risky nature of agricultural production and marketing associated with the production environment. As the literature shows, farm assessments of the impact of Bt cotton in China, India, and South Africa indicate that there is considerable variability in economic returns (Fok, Liang, and Wu 2005; Bennett et al. 2006; Qaim et al. 2006). Advanced risk modeling not only deals with this variability in returns but it also helps to explain the behavior of risk-averse actors, such as the producers in developing countries.[3]

The approach adopted in this study—economic surplus combined with advanced risk modeling—also allows us to deal with two other important issues. The first one is related to the institutional arrangement for the approval and adoption of the GM technology. Analyses conducted in South Africa (Gouse et al. 2005), Mexico (Traxler et al. 2003), and Argentina (Qaim and de Janvry 2005) confirm the importance of institutional arrangements in determining the economic benefits earned by farmers. Institutional factors include, for example, the nature of producer contracts, the magnitude of the technology fee, the availability of credit, and the extent of competition in the seed and product markets.

---

3 One can derive supply and demand functions under more general conditions and assumptions that do not necessarily assume profit maximization, as in the Alston, Norton, and Pardey (1995) economic surplus approach. Further, derived supply and demand functions can incorporate risk considerations. However, the derivation of such functions requires data that are not necessarily available in Uganda and other developing countries, especially not when doing an ex ante assessment that might be included in a biosafety or technology regulatory process used to decide on whether to approve the technology. In this situation, there are several limitations that assessors in developing countries will face, including time, financial resources, and human capacity to conduct such assessments. We thus chose the economic surplus approach as delineated by Alston, Norton, and Pardey, which is probably the most parsimonious in terms of data and which has a long history of use in ex ante evaluations. This approach can also be expanded and modified to include more sophisticated features, such as the use of stochastic parameters and real-option approaches.

The second issue is the damage-abatement effect of the GM technology.[4] In essence, the expected value to farmers of a trait such as insect resistance (as found in Bt cotton), herbicide tolerance (as in herbicide-tolerant [HT] cotton), or both (as in a stacked cotton variety) is directly related to the presence and intensity of pest or weed infestations.[5] If no pest or weed infestation exists, there is no particular reason GM cotton would yield more or have a different production and cost structure than conventional cotton, assuming other productivity traits (such as germplasm) are the same.[6] By ignoring the damage-abatement effect—and therefore not determining how much of a pest infestation there is to be reduced—economic surplus models could overestimate the benefits of insect-resistant technologies (Pemsl, Waibel, and Gutierrez 2005).

In the standard economic surplus model, the estimated magnitude and distribution of the economic benefits depend on many factors. These include—but are not limited to—(1) the price elasticities of crop supply and demand; (2) the volume of production and whether the country is a large or small producer; (3) trade issues, such as whether the country exports or imports the crop; (4) the nature of the innovative change induced by the technology; (5) the uniqueness of crop attributes and traits; and (6) the relevance of traits for genetic enhancement, such as agronomic traits and traits for resistance to extreme weather conditions, pest infestations, or both.[7] Data are typically drawn from some combination of sources, including sample surveys of farmers, trial data (field and greenhouse), secondary data, or all three. The analysis can be conducted at the regional, national, or global level.

The basic Alston, Norton, and Pardey (1995) model is based on the assumption that the adopting country's markets and economy, as well as its technological adoption pattern, can be modeled by supply, demand, and market equations. Linear supply and demand functions are used, because they do not need much data for estimation and they facilitate calculating the size of

---

4   Damage abatement is defined as the proportion of the destructive capacity of the pest (the "damaging agent") eliminated by the application of a given level of chemical or physical control (the "control input").

5   A stacked crop variety is defined as one that has two distinct traits, such as the resistance to bollworm attacks and the tolerance to herbicide applications. Stacked cotton that is both pest resistant and herbicide tolerant is known as Bt/RR cotton.

6   In fact, in the case of GM crops, farmers face a scenario in which they may pay for the technology in advance without knowing whether there will be a pest infestation. Farmers in this situation may incur additional financial risk, as they may pay for a technology upfront that they may not need in that particular crop cycle. Of course, this issue enters into the realm of the insurance value of using such technologies.

7   A large producer is one that would set international prices (a price setter), whereas a small producer is one whose production does not affect international prices (a price taker).

consumer surplus changes. Besides linear functions, other functional forms have been proposed in the literature, such as supply and demand functions with constant elasticities (Ayer and Schuh 1972; Scobie and Posada 1978), kinked supply functions (Rose 1980), and a constant elasticity form with a positive price intercept (Pachico, Lynam, and Jones 1987). Alston, Norton, and Pardey (1995) indicate that the type of function is not as important as the nature of the supply shifts and the elasticities used in the estimation procedure.

The nature of the supply can have significant implications on the magnitude of impacts. The new supply function resulting from technological change can be parallel to the old one representing the state of nature before the change. Alternatively, the new supply function can have a different slope with respect to the original function, becoming either convergent with or divergent from the original function. For example, with linear supply and demand functions, a parallel shift of the supply function will yield twice the total benefits of equivalent pivotal shifts with respect to the pre-innovation equilibrium. Further, with a parallel shift, producers always benefit from innovation, unless the supply function is perfectly elastic.[8]

We concur with Alston, Norton, and Pardey (1995) that in the absence of detailed information about the supply and demand functions, it is wise to choose linear functions with parallel shifts. A linear supply and demand function with parallel shifts is the simplest starting point from which to derive a supply-and-demand system, because such a function relies on a single data point to describe the entire system. The single data point used is the cost change induced by the use of the new technology, which can be used to map the linear supply and demand functions (Rose 1980). Assuming linear supply and demand functions also provides the additional advantage of producing results that can be compared to a rich history of economic surplus estimations based on the same set of assumptions to build a model (Alston, Norton, and Pardey 1995; for example, see Norton and Hautea 2009).[9]

---

8 There is no difference between the actual price the producer receives for the commodity and the minimum price he is willing to accept.

9 Demand functions can be derived from utility functions. This approach allows incorporation of risk preferences and other interesting characteristics when modeling technology adoption. Similarly, supply functions can be derived from appropriate data. We chose not to pursue this approach, as it requires more data than what was (and may be) readily available in developing countries for estimating the benefits of innovation from technology adoption. This approach also requires knowledge and expertise to help sort out many of the methodological hurdles dealing with the estimation of supply and demand curves. The economic surplus model should in any case be a rapid assessment and perhaps first approximate the true benefits from technology adoption in developing countries.

The Alston, Norton, and Pardey (1995) approach is described by the following system:

Supply: $$Q_S = \alpha + \beta\,(P + k) = (\alpha + \beta k) + \beta P \tag{6.1}$$

Demand: $$Q_D = \gamma - \delta P \tag{6.2}$$

Market clearing: $$Q_S = Q_D. \tag{6.3}$$

In this system—and throughout this chapter—Q_S is the quantity supplied, Q_D is the quantity demanded, k is the shift in supply stemming from the introduction of the innovation, and P is the equilibrium price. Variables α and γ are intercepts, whereas β and δ are the slopes of the supply and demand equations. A graphic representation of this model is presented in Figure 6.1.

The relative reduction in price (Z) is defined as

$$Z = \varepsilon K/[\varepsilon + \eta] = -(P_1 - P_0)/P_0, \tag{6.4}$$

where ε is the elasticity of supply, η is the absolute value of elasticity of demand, P_0 is the price before innovation adoption, P_1 is the price after innovation adoption, Q_0 is the quantity before innovation adoption, and Q_1 is the quantity after the innovation adoption. We define K as the relative change in price compared to the pre-innovation price (P_0): $K = k/P_0$.

After setting $Q_S = Q_D = Q$, the general formula to estimate price is $P = (\gamma - \alpha - \beta k)/(\beta + \delta)$. If $k = 0$, then $P_0 = (\gamma - \alpha)/(\beta + \delta)$; if $k = KP_0$, then $P_1 = (\gamma - \alpha - \beta KP_0)/(\beta + \delta)$.

Therefore, the change in price is $P_1 - P_0 = -\beta KP_0/(\beta + \delta)$. The absolute value of the relative change in price (Z) is $-(P_1 - P_0)/P_0 = \beta KP_0/(\beta + \delta)$. Multiplying the numerator and the denominator by P_0/Q_0 and manipulating algebraically yields the equivalent elasticity formula (6.4)

For the closed economy model, algebraic formulas for producer, consumer, and total surplus are

Change in consumer surplus: $$\Delta CS = P_0 Q_0 Z(1 + 0.5Z\eta) \tag{6.5}$$

Change in producer surplus: $$\Delta PS = P_0 Q_0 (K - Z)(1 + 0.5Z\eta) \tag{6.6}$$

Change in total surplus: $$\Delta TS = \Delta CS + \Delta PS = P_0 Q_0 Z(1 + 0.5Z\eta). \tag{6.7}$$

FIGURE 6.1 The general closed economy model for economic surplus

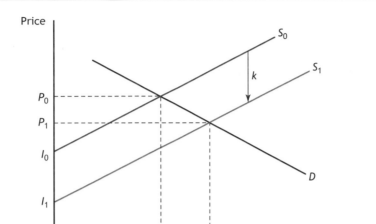

Source: Authors.

Note: D = demand function; k = shift in supply stemming from the introduction of the innovation; P_0 = price before in-novation adoption; P_1 = price after innovation adoption; Q_0 = quantity produced before innovation adoption; Q_1 = quantity produced after innovation adoption; S_0 = supply function before innovation adoption; S_1 = supply function after innovation adoption.

These formulas estimate the change in consumer, producer, and total surplus stemming from the adoption and use of a GM innovation. The values estimated are the additional values produced by the technology in the hands of farmers.

In the case of the small open economy (Figure 6.2 presents the case of a net exporter country), the change in consumer surplus is equal to zero ($\Delta CS = 0$), as Ugandan farmers are price takers (that is, they cannot influence world prices because of their small-scale production and so must sell at whatever world prices are); thus, the reference price is the world price. Consumers in Uganda do not benefit, as there is no price reduction from the adoption of the innovation. Therefore, producer surplus equals total surplus ($\Delta PS = \Delta TS$).

The formula for producer surplus in the small open economy model can be estimated by taking the mathematical limit of the formula for change in producer surplus when the demand elasticity approaches infinity (Alston, Norton, and Pardey 1995). Therefore, the formula for producer surplus in the small open economy model is

$$\Delta PS = \Delta TS = P_w Q_0 K(1 + 0.5Z\varepsilon), \tag{6.8}$$

FIGURE 6.2   Supply and demand for a small open economy: The case of the net exporter

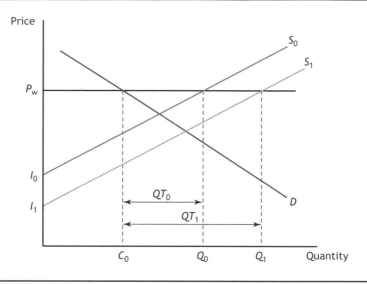

Source: Authors.

Notes: C_0 = quantity demanded at price P_w; D = demand function; P_w = world price; Q_0 = quantity produced before innovation adoption; Q_1 = quantity produced after innovation adoption; QT_0 = quantity traded before innovation adoption; QT_1 = quantity traded after innovation adoption; S_0 = supply function before innovation adoption; S_1 = supply function after innovation adoption.

where ΔPS is the change in producer surplus, ΔTS is the change in total surplus stemming from the introduction of an innovation, Q_0 is the quantity produced before innovation adoption, and P_w is the world price of cotton.

The empirical formulas for shifts in the supply curve (K), changes in the producer surplus (ΔPS), and the change in annual net benefits (ΔNB) used in the study are

$$K = \left[\frac{\Delta Y}{\varepsilon_2} + \frac{\Delta C}{(1 + \Delta Y)} - \frac{C_{TF}}{TC} \right] \times A \tag{6.9}$$

$$\Delta CS = 0 \tag{6.10}$$

$$\Delta NB = \Delta PS - C_{RD+Reg} - C_{TF}, \tag{6.11}$$

where K is the proportional size of the supply shift, ΔY is the yield difference expressed as the expected yield difference between GM and conventional cotton, ε_A is the elasticity of total cotton supply, ΔC is the production cost difference between GM and conventional cotton, C_{TF} is the technology fee, TC is the total costs of production, and A is the adoption rate. C_{RD+Reg} represents the costs necessary to develop and deploy the technology, including biosafety regulatory compliance costs and adaptive research and development (R&D).

Expected yield differences are converted to equivalent cost changes by dividing the percent yield difference (ΔY) by the elasticity of supply (ε_A). Production cost changes (ΔC) stemming from the introduction of the technology are converted to equivalent cost changes by dividing by 1 + percentage yield difference ($1 + \Delta Y$).

As described in Alston, Norton, and Pardey (1995), the present value of net benefits and the internal rate of return (IRR) to society from the potential use of GM cotton in Uganda are calculated from the stream of yearly changes in producer surplus minus the investment necessary to bring the technology to farmers. In this case, such investments as R&D, regulatory costs, and technology fees are included. The annual net benefit changes (ΔNB), calculated over the years of the expected adoption time period, are discounted to present value by dividing by $(1 + i)^t$, where i is the discount rate and t is the number of years after adoption started in Uganda.

Limitations of the Economic Surplus Approach

Alston, Norton, and Pardey (1995) and other authors have pinpointed the advantages and limitations of the economic surplus approach. The major advantages are that this approach is parsimonious with respect to data and can be used to portray the distributional effects of various institutional and market structures. The principal disadvantages are that consumer surplus is Marshallian; thus, this measure does not consider changes in income. Farmer prices and quantities of other commodities are fixed, and impacts on input markets are not explicitly considered. Furthermore, farmers are considered to be risk neutral—indifferent to risk and thus willing to consider changes only in expected output. Further, year and location variations are not included in estimations unless explicitly included. Thus, standard models do not consider heterogeneity, which can lead to bias in estimations.

The homogeneity assumption is probably the most challenging limitation for the case of cotton production in Uganda. Producer heterogeneity, however, is an issue not often addressed in ex ante studies. As Demont et al. (2008) have discussed, in most ex ante studies, the need exists to estimate potential benefits of technology adoption on the basis of information that is incomplete, imperfect, or both. Studies that do not explicitly include individual farmers and spatial heterogeneity run the risk of having significant homogeneity bias in their results. For example, the decision to adopt a GM crop (such as Bt cotton) will be greatly influenced by, among other factors, the level and frequency of pest infestations that caused economic damage. These factors vary significantly across farmers and locations, thus introducing heterogeneity into the process.

The issue of producer heterogeneity is even more complicated when an innovation, such as Bt cotton, is deployed in an imperfectly competitive seed market. In a standard monopoly situation and under a defined set of conditions, a monopolistic innovator with a technology deemed to be sufficiently different from existing technologies—a "drastic innovation"—may capture most if not all rents derived from the technology's adoption. However, if the set of potential adopting farmers is indeed heterogeneous, an innovator who has a monopoly over the innovation may not capture all rents even if the technology is a drastic innovation. In these cases where a monopolistic innovator does not capture all rents, sometimes greater benefits can be captured by adopting farmers (Alexander and Goodhue 2002; Oehmke and Wolf 2004).

The method proposed by Demont et al. (2008) to address both Bt and HT traits in an ex ante setting still requires information about the distribution of expenditures on herbicides and insecticides. This information needs to be disaggregated by region and by target pest to reflect a specific country's producer heterogeneity. This information is not readily available in Uganda, although some secondary information about pesticide expenditures could be gleaned from the household surveys conducted in the country. In our estimations of the potential impacts from GM cotton adoption in Uganda, we use probability distributions that describe both yield losses and insect-control costs stemming from the target lepidopteran insects and weeds.

The probability distribution used in this study to simulate potential gains from GM cotton adoption is based on the farmer survey we conducted in Lira and Kasese. We aggregated the farmer survey results to create a national probability distribution covering a wide set of potential outcomes in the simulation. The simulation is repeated for an ample number of iterations to ensure a stable outcome. This approach partially addresses spatial heterogeneity in our economywide estimates of impact. The limited scope of our survey introduced some limitations on trying to extrapolate to the country as a whole. In other words, our results need to be viewed as a first approximation to the estimation of impacts for this technology.

The quality of the underlying data is crucial to the validity of the results. In general, reliable cross-sectional time series data for these technologies are not yet available in developing economies, because they are too costly to obtain. In contrast, in the United States extensive surveys are conducted continually (for example, the USDA Agriculture Resource Management Survey, on which many detailed analyses have been based), and cheaper methods (mail and phone interviews) are feasible.

Two points should be emphasized. First, adapting standard economic models (which represent markets of different sizes and degrees of openness) by explicit modifications of the structural equations in the models is an effective way to treat some of the methodological challenges described above. Second, this methodology provides the type of information that most national policy-makers and investors in technology development consider to be fundamental to the decisionmaking process.

The assumptions used to build the standard economic surplus models perhaps best reflect an industry with commercially oriented farmers who buy and sell in well-organized markets and who grow their crops under relatively homogeneous growing conditions. Nevertheless, the economic surplus model can be used judiciously for a developing country if its limitations are fully understood.

Methodology

The two methods used in this chapter—economic surplus and stochastic dominance—are implemented to answer the following research questions: What are the net benefits and risks associated with the adoption of Bt and HT cotton in Uganda? What institutional arrangements for technology delivery would potentially generate more benefits for the society? Both methods rely on parameters and on key assumptions. The assumptions are mainly used to assign values to the parameters. Furthermore, the estimations are done for six different scenarios.

Methods

STOCHASTIC ECONOMIC SURPLUS

The study implements an extension to the standard economic surplus model—as described in Alston, Norton, and Pardey (1995)—and incorporates stochastic elements. The conventional economic surplus approach is also termed a *partial equilibrium displacement model,* because it considers the effects of the technology change only in the market where the change occurs. Effects in other markets, such as input markets, are disregarded. Because Uganda is a small cotton producer and a price taker, we used the small open economy model here.

The standard model is enhanced in the study to account for risk, uncertainty, and sparse data by replacing single-point values with probability distributions for selected parameters (technology fee, supply elasticity, yield, and cost differences between Bt and non-Bt varieties). The enhanced economic surplus model provides a more rigorous sensitivity analysis of model parameters (Davis and Espinoza 1998; Falck-Zepeda, Traxler, and Nelson 2000; Zhao et al. 2000; Fisher, Masters, and Sidibe 2001; Hareau, Mills, and Norton 2006).

For each of the stochastic parameters the study pursued different approaches, depending on available data, to define the probability distribution. In some cases, distributions for the data collected in the field surveys were estimated using the "Best Fit" routine included in the @Risk software (Palisade Corporation 2012). In other cases, we used triangular distributions based on values cited in the published literature. Triangular distributions are widely used in the analysis of agricultural risk and uncertainty. The triangular distribution is a continuous probability distribution that consists of the minimum, maximum, and mode. For a large number of observations, this triangular distribution approximates the normal distribution (Hardaker et al. 2004).

The @Risk program calculates and saves designated output variable values from repeated sampling processes of the specified triangular distributions. After repeated sampling, the average variation across all simulations will tend to decrease. Thus, a recommended practice while running simulations is to monitor the convergence to a stable steady state for selected output variables. On the one hand, practitioners need to ensure that sufficient iterations are run so that the statistics describing all the iterations are deemed to be stable and therefore reliable. On the other hand, after a certain point, additional iterations will not yield new information, because there is little variation in the average statistics that summarize all the iterations,

Our study used a conservative approach and thus ensured convergence to a stable state but performed at least 10,000 iterations per simulation. This approach may be redundant in some cases but serves the purpose of describing output variables in finer detail. Output variables in the estimations include producer surplus, consumer surplus, total surplus, present value of net benefits, and IRR. Few studies have applied a stochastic approach, and thus our study further advances the utility of this approach for empirical policy analysis (Pemsl, Waibel, and Orphal 2004; Hareau, Mills, and Norton 2006; Falck-Zepeda, Horna, and Smale 2008a).

We performed an advanced sensitivity analysis using @Risk. In an advanced sensitivity analysis, the values of the probability distribution of parameters deemed interesting are allowed to increase or decrease on a percentage basis. The program then runs the set number of iterations while calculating the output variables of interest. In our case, we performed an advanced sensitivity analysis on cotton prices and the technology fee by allowing percentage increases that went from 0 percent (baseline) to 70 percent by performing 10 percent step increases.

Changes in adoption rates over time were modeled using a conventional s-shaped adoption curve. The maximum adoption rate was elicited from cotton experts consulted in Uganda. Data from the International Cotton Advisory Committee surveys (ICAC 2008) described production cost and cotton prices.

It is important to note that one can construct economic surplus estimates that can be disaggregated to the level of administrative or agroecological zone as needed. This ability to disaggregate economic surplus estimates allows the estimation of producer and consumer surpluses by regionally defined subgroups, as was done for Bt cotton in the United States (see Falck-Zepeda, Traxler, and Nelson 2000). The only requirement to implement this alternative is to include in the empirical formula a probability distribution for each of the administrative units or agroecological zones in the country and then weigh each distribution by its relative share of national production.

The aggregation of the individual administrative units or agroecological zones can be used to derive an aggregate national supply shift through the K variable, which can be used in the estimation of producer surplus as in equation (6.8) (see Falck-Zepeda, Traxler, and Nelson 2000). If the region, district, or both represent a relatively small share of cotton production, the unit or zone's impact will also be small. As information at this level is not available for Uganda, we pursue the strategy of deriving a national estimate for K. The assumption involved in deriving this estimate is that all regions in Uganda have a similar yield and cost probability distribution. This assumption is limiting, but it is the best that can be done with the available information for Uganda.

STOCHASTIC DOMINANCE

Chapter 5 describes the stochastic dominance method; additional details are presented in Appendix 6. We used the results from the economic surplus model (10,000 iterations for each of the six scenarios described below) and ran a robust stochastic dominance analysis using the computer program SIMETAR (©2008, Simetar Inc., College Station, Texas).

Model and Parameters

TECHNOLOGY ADOPTION CHARACTERISTICS

The total timeline for GM cotton use, including R&D, regulatory time lags before adoption, adoption, and eventual disadoption, is 25 years. It includes 5 years for initial adoption, 10 years of effective product lifespan, and 5 years for disadoption. This may seem like a long period, but it reflects the global experience with Bt/HT cotton technologies. For instance, farmers in the United States still planted the variety NuCOTN 35B, one of the two GM varieties initially released in the United States in 1988, in 2008. NuCOTN 35B was a dominant variety for 5–7 years after its introduction. Yet the current planting rate of NuCOTN 35B is a small percentage (less than 0.5 percent) of total area planted to cotton.

The ceiling for the adoption curve is the best estimate, based on historical levels of damage and geographical dispersion, made by cotton experts in Uganda. The counterfactual in all cases is the continued use of Bukalasa Pedigree Albar (BPA), the local variety of cotton. Changes in producer surplus, measured as changes in net benefits, are thus relative to the counterfactual, which is the conventional cotton variety.

Because of the ex ante character of this study, no complete data are available to estimate the cost of R&D and biosafety regulatory compliance. Therefore, available data from other developing countries were used, including that from India and China (Falck-Zepeda and Cohen 2003; Pray et al. 2006). Notice that when the cost of compliance with biosafety regulations is included, producer benefits, all else being equal, decrease. This decrease may be offset by reductions in yield damage and pesticide costs, however.

STOCHASTIC PARAMETERS

The parameters used in the economic surplus model, along with the assumptions made and literature consulted to obtain parameter values, are presented in Tables 6.1 and 6.2. Table 6.1 introduces those parameters where we substituted a static mean value for a probability distribution. Table 6.2 includes baseline parameters used in the model.

Next we discuss only those parameters that are critical to the model and may be controversial.

1. *Technology fee.* The technology fee follows a triangular distribution with a minimum value of 15, most likely (mode) value of 32, and a maximum of 56. These values correspond to the values for other countries, including South Africa, found in the literature. In the new distribution used for the

TABLE 6.1   Stochastic distributions of parameters included in
economic surplus model simulations

Parameter	Mean value	Stochastic distribution	Justification
Technology fee for GM crops (US$ per hectare)			
Technology fee	24.00	Triangular distribution with minimum = 15, mode = 32, maximum = 56	Range of values found in the literature (Falck-Zepeda, Traxler, and Nelson 2000; Huang et al. 2003, 2004; Bennett et al. 2004): 15 corresponds to India, 32 to South Africa and China, and 56 to the United States
Reduced technology fee	11.44	Triangular distribution with minimum = 5, mode = 10.66, maximum = 18.66	Estimated range found by reducing technology fees from the literature to 33 percent of their original value
Elasticity of cotton supply	0.93	Triangular distribution with minimum = 0.3, mode = 1, maximum = 1.5	Dercon (1993); Alston, Norton, and Pardey (1995); Delgado and Minot (2000); Minot and Daniels (2005)
Cotton yield difference (percent)	≈19.6	Logistic distribution with $\alpha = 0.0939$, $\alpha = 0.168$, truncated at −10 and 74 percent	Estimated using @Risk software on the household survey for existing cotton yields; truncation is the maximum yield estimate from the survey
Cotton cost differences			
Difference stemming from GM crops (reduction in pesticides/spraying) (percent)	13.00	Triangular distribution with minimum = −1, mode = 3, maximum = 16	Mean and maximum values are estimated from the household survey; minimum value is an estimated average control cost for lepidopterans relative to total costs in the survey
Difference in number of glyphosate applications	1–2	Uniform distribution with values between 1 and 2 applications	Literature review and experts consulted during project
Producer price (US$ per kilogram)	0.188	Triangular distribution with minimum = 0.06, mode = 0.12, maximum = 0.26	Tschirley, Poulton, and Labaste (2009) and the household survey

Source: Authors.
Note: Mean values = the average value of the distribution described in the stochastic distribution column in this table.

economic surplus model, we assumed a reduced technology fee representing 33 percent of each of the original values.

2. *Elasticity of cotton supply.* Given the limited information concerning the supply elasticity of cotton in Uganda (and in Africa in general), unitary elasticity was assumed to be the most likely elasticity. To set the range of the supply elasticity values, including both unitary and other elasticity values, the literature was consulted. The final triangular distribution used took a minimum value of 0.3, a maximum of 1.5, and a mode value of 1.0.

TABLE 6.2   Baseline data used for economic surplus model

Parameter	Value	Source
Total cotton area in 2007 (hectares)	195,000	FAO (2010)
Total area planted with organic cotton (hectares)	40,000	Reports on area varied from 8,900 hectares in 2001 (Willert and Yussefi 2007) to 122,000 hectares (FAO 2010)
Total area planted with conventional cotton (hectares)	155,000	Estimated by subtracting estimate of total area planted with organic cotton from total cotton area in 2007
Seed cotton quantity produced (kilograms)	117,587,916	Estimated by multiplying area by yield per hectare
Total value of production (US$)	7,038,712	Estimated by multiplying quantity produced by output price
Average value of production (US$ per hectare)	114	Estimated by dividing total value of production by total hectares of cotton
Average cost of production (US$ per hectare)	241	Average of 151 respondents' answers to the household survey
Total (national) cost of production (US$)	14,925,567	Estimated by multiplying average cost of production by total hectares of cotton
Total (national) net value (US$)	(−7,886,856)	Estimated by subtracting total cost of production from total value of production

Source: Authors.

3. *Cotton yield difference.* The study drew values for the distribution of cotton yields from the 150 farmers included in the household survey (see Chapter 2). As there is no adoption in Uganda, the study used data obtained from the survey on cotton yield losses induced by the target pest. Yield differences between the stacked Bt/HT cotton variety and conventional cotton were elicited from experts and from the experiences described in the literature, but they were adjusted to reflect losses to weeds and lepidopteran insects endured by Ugandan farmers. These yield-difference values were used as proxies to evaluate the potential damage-abatement effect of the technology. Note that the minimum loss-abatement value of 0 percent allows for the possibility of no yield difference being induced by the Bt/HT traits. If there is no pest or weed infestation, there is no reason to observe a yield difference between the Bt/HT and the conventional variety. If a negative yield difference—the conventional variety yields more than the Bt/HT variety—is observed, it is most likely unrelated to the trait but rather the result of the germplasm used. To capture variation in the simulations over time, a probability distribution of yield differences was imputed for each year of the simulation. One could expect that as time

progresses, the impact of a particular year on the present value of net benefits would decline.

4. *Cotton cost difference.* The adoption of Bt/HT cotton could induce a per unit cost savings from reduced pesticide use. As in the case of yield differences, our study used the household survey data—in this case, to determine the cost of controlling lepidopteran insects—and assumed that the technology would eliminate the need for these pesticide applications. The simulations included values based on the household survey to produce a triangular distribution of pesticide applications and costs. Farmers were asked about the amounts and costs of pesticide used against Lepidoptera, the insect family that bollworm belongs to. Bt/HT cotton has an effect mainly on bollworm but also on other insects from this family. The cost advantage was then defined as the cost difference given the lower use of pesticides that control lepidopterans. The responses across farmers provided information to set the minimum, maximum, and mode values that define a triangular distribution.[10]

The minimum cost difference was set to −1 percent to account for cases where the Bt/HT variety does not reduce insecticide applications— perhaps because of no insect pressure—while the farmer still has to pay the additional cost of the Bt/HT cottonseed. Setting the minimum cost difference to a negative value implies that there is a cost implication in adopting Bt/HT cotton when there is no pest attack. To get a more precise estimation of the cost to farmers when there is no pest attack requires having a better understanding of the probability distribution of pest attacks, which is not available for Uganda. We pursue the option of setting the minimum cost difference to the value of the loss of the technology fee only.

The experience of other countries has shown that even in the case of successful adoption of a Bt/HT cotton technology and successful control of the primary pests, secondary pest populations can become economically important and therefore may require pesticide applications. The cost of controlling secondary pests then could offset benefits from reduced applications of pesticides to control the primary pests. Notice that the

---

10 Demont et al. (2008) propose that the triangular distribution introduces a heavier emphasis on the tails of the distribution than what experience has shown to be the case in practice. These authors suggest using a Pert probability distribution function (derived from the more general beta function), because it is also parsimonious and does not emphasize the distribution's tails as the triangular distribution can. A Pert probability distribution requires knowledge about the cost structure under assessment, however. In lieu of more information about the herbicide cost structure in Uganda, we chose to use the triangular form.

simulations account for the possibility that farmers may be worse off by using the Bt/HT cotton technology when there is no yield advantage, no costs advantage, or both.

In the case of herbicide use, the literature and experts were consulted to find the rate of technical exchange between chemical and manual weeding. To simulate the use of herbicide, estimations done in the study substitute manual weedings for herbicide applications. We used a rule of thumb that one herbicide application eliminates three manual weedings. In simulations, it was assumed that the rate of herbicide application follows a uniform distribution of either one or two applications. Thus, between three and six weedings may be eliminated by using herbicide. As weeds are usually more prevalent than insects, the study did not assume a zero application of herbicides. If in fact farmers have no access to herbicides, this assumption will result in an overestimation of benefits.

5. *Producer price.* We use a triangular distribution defined with a minimum value of $0.06 per kilogram, mode value of $0.12 per kilogram, and maximum value of $0.26 per kilogram.[11] These values were derived from the household survey and a secondary source.

6. *Discount/interest rate.* Although not a stochastic parameter, the discount rate is still an important part of the economic surplus model. Among the issues to be considered are which discount rate and which form of the discount rate distribution (that is, logistic or hyperbolic) should be used to discount annual cash flows to bring them to present values in investment projects. What the economic surplus method does is estimate total social returns that can be disaggregated to producer, consumer, and innovator shares. Thus, we were interested in using discount rates that reflect society's preferences for holdings now rather than in the future (the time value of money) and attitudes toward risk. The range of discount rates used for public sector projects ranges from 6 to 9 percent (for example, New Zealand uses a 10 percent discount rate; see Young 2002). The consensus in the literature is that investment projects that can be classified as higher risk than other projects should use a higher interest rate for discounting future cash flows produced by the investment, to reflect this risk. For our study, the interest rate in 2008 of 18 percent was chosen to reflect the higher risk of investments needed to transfer the Bt/HT cotton technology to Uganda. Note that in a study by Kikulwe (2010) estimating the

---

11 All dollars are US dollars in this chapter.

benefits from the potential adoption of a GM banana in Uganda, he uses a discount rate of 12 percent but chooses to vary this discount rate from 0 to 16 percent to measure the sensitivity of his results to the discount rate factor.

Scenarios: Description and Simulations

The scenarios described in this section compare the current situation (the use of the conventional cotton variety BPA) in Uganda to the use of Bt/HT cotton. We used the described adoption pathway over the whole cotton area in Uganda. Table 6.3 presents those parameters that are the starting assumptions for the counterfactual of continued conventional cotton use and that will change because of the adoption of Bt/HT cotton in Uganda. In all scenarios, the developer invests resources in obtaining regulatory approval; thus, these costs are included in the estimations.

SCENARIO I: PUBLIC SECTOR RELEASE

This is the benchmark scenario, in which the public sector releases a Bt/HT cotton variety using the same price as the conventional BPA variety. Currently, seed is given to farmers free of charge. Thus, this scenario assumes that there will be no price differential between the Bt/HT variety and the conventional

TABLE 6.3   Assumed baseline data used for economic and cost parameters

Parameter	Value	Source
Weeding costs for hired labor (US$ per hectare)	16.49	Average for survey respondents
Base year	2008	
Interest rate (percent)	18	Bank of Uganda
Exchange rate, 2006 (Ugandan shilling/US$)	1,710	Bank of Uganda
Elasticity of demand	0.50	Literature review
R&D and regulatory delay (years)	5	Experts consulted for project
Adoption (time to maximize use) (years)	5	Experts consulted for project
Time of effective product lifespan (years)	10	Experts consulted for project
Time for disadoption (years)	5	Experts consulted for project
Total time of use (years)	25	Experts consulted for project
Maximum adoption level (percent)	40	Experts consulted for project
Total R&D and regulatory costs for GM crops (US$)	300,000	Falck-Zepeda and Cohen (2003); Quemada (2003); Pray et al. (2006)

Source: Authors.
Note: GM = genetically modified; R&D = research and development.

variety. The underlying assumption is that farmers manage only the insect-protection aspect of the technology or that the variety released to farmers incorporates only insect protection, not the herbicide trait. This assumption is relaxed in scenarios IV and V.

SCENARIO II: PRIVATE SECTOR RELEASE WITH TECHNOLOGY FEE

The private developer (or the public sector) releases the Bt/HT cotton technology but charges a technology fee similar to those levels charged in other developing countries. As a result, Bt/HT cotton costs more than the conventional BPA variety.

SCENARIO III: PRIVATE SECTOR RELEASE WITH REDUCED TECHNOLOGY FEE

The private developer charges a technology fee at a reduced level compared to scenario II. The reduced fee corresponds to a reduction of two-thirds relative to the price charged in scenario II. The objective of this scenario is to measure the impact of negotiating a lower technology fee with the developer and to highlight the effect of reducing input costs for producers. Note that scenarios I–III correspond to a situation in which farmers obtain access to a variety with only the Bt trait or are interested in only the Bt component of a stacked-gene Bt/RR variety.[12] This process allows separating the effect of insect resistance and herbicide tolerance traits.

SCENARIO IV: DUAL INSECT RESISTANCE AND HERBICIDE TOLERANCE WITH TECHNOLOGY FEE

Farmers receive and use a cotton variety with both the insect resistance and herbicide tolerance traits and are charged a technology fee at the level charged internationally (as in scenario II). Farmers apply the herbicide glyphosate at a rate that varies stochastically between one and two applications; their reduction in manual weeding fluctuates between three and six cycles of weeding. Farmers' ability to apply the herbicide glyphosate, which is subject to their access to credit, the herbicide's physical availability, or both, is critical in this scenario. As described above, few farmers in the survey use herbicides, much less glyphosate. This low herbicide use and the negligible benefits from herbicide tolerance that result from it are policy issues that developers and policymakers in Uganda will need to address to ensure capturing the potential benefits of the technology.

---

12 It is possible that, even though herbicide applications do not seem to be common practice in Uganda, farmers could still use a stacked variety for its advantage of controlling bollworm and would manage the crop accordingly.

**SCENARIO V: DUAL INSECT RESISTANCE AND HERBICIDE TOLERANCE TRAITS
WITH REDUCED TECHNOLOGY FEE**

This scenario uses the same assumptions as scenario IV, except for a reduced
technology fee corresponding to one-third the price charged internationally
for the technology.

**SCENARIO VI: DUAL TRAITS AND REDUCED TECHNOLOGY FEE,
WITH INCREASED TOTAL COTTON AREA PLANTED AND ADOPTION RATE**

This scenario uses the same assumptions as scenario V but doubles the
maximum cotton area from the current average of 195,000 hectares to
380,000 hectares and doubles the maximum adoption rate from the baseline of 40 percent to 80 percent. The purpose of this scenario is to highlight
the differences in impact that can be attributed to increased areas and higher
adoption rates. Land expansion to increase cotton production is likely to trig-
ger a set of additional costs to individual farmers and to society. These may
include the costs needed to clear land and the decreases in productivity resulting from bringing marginal land into production.

To estimate all the additional costs and benefits for other sectors of the economy, one would need to have an economic model that incorporates the multiple supply and demand interactions, and such factors as use of labor and
availability of inputs and capital. Such estimations can be done only with
economywide models, such as the CGE model for Uganda.

In the absence of a CGE model, one could introduce much more detail
into the supply and demand functions in an economic surplus model to estimate some of the costs and benefits of scenario VI. That economic surplus
models are partial equilibrium models will severely limit this approach, however. As there is no readily available CGE model for Uganda, or additional
data to introduce the interactions with other sectors of the economy into the
economic surplus model, we acknowledge that estimates derived from scenario
VI may be incorrect, as they may overestimate the level of gains (or alternatively, underestimate the losses).

Results

The results of the simulations show that changes in total net benefits
(Table 6.4) vary significantly from scenario to scenario. Moreover, the introduction of a technology fee significantly reduces the level of benefits to farmers (Table 6.4 and Appendixes 8–10). In scenario I, with a free public sector

TABLE 6.4   Summary of economic surplus results

Scenario	Description	Present value of net benefits (US dollars)	Total nominal benefits (US dollars)	IRR[a] (percent)
I	Public-sector release	6,804,135 (−101,987 to 17,235,159)	64,997,970 (724,666 to 161,818,275)	81 (43–117)
II	Private-sector release with technology fee	1,533,175 (−1,856,470 to 7,969,868)	15,989,645 (−15,041,323 to 73,167,634)	54 (10–99)
III	Private-sector release with reduced technology fee	4,121,746 (−633,840 to 11,827,255)	39,961,923 (−41,154,973 to 109,740,933)	68 (30–106)
IV	Both insect resistance and herbicide tolerance with technology fee	1,821,044 (−1,846,490 to 8,461,933)	18,626,829 (−15,087,545 to 78,786,758)	56 (11–99)
V	Both insect resistance and herbicide tolerance with reduced technology fee	7,473,349 (−1,334,243 to 21,043,419)	71,299,556 (−10,638,492 to 194,953,800)	83 (38–125)
VI	Dual trait with increased total cotton area planted and adoption rate	32,242,210 (−9,603,559 to 129,156,000)	297,960,600 (−85,966,700 to 1,189,203,000)	131 (52.8–211)

Source: Authors.

Notes: Values in parentheses are the 5 percent and 95 percent percentiles, which can be construed as the confidence interval for each mean value for the 10,000 simulations in each scenario. Total nominal benefits are the sum of all annual (net) benefits in nominal terms (without discounting). IRR = internal rate of return.

[a]IRR is calculated by @Risk as the discount rate that would make the present value of all annual cash flows equal to zero. The estimated IRR is used to compare the investment to the interest rate describing the cost of capital.

release, the present value of net benefits during the 25 years of the simulation is $6,804,135, as opposed to $1,533,175 in scenario II, which includes a technology fee. Decreasing the technology fee to one-third of what is charged internationally increases present value of net benefits to $4,121,746 in scenario III. This result highlights the importance for countries or individuals to negotiate the technology fee level. A low technology fee increases the likelihood that farmers will gain from the introduction of the technology.

The introduction of the herbicide trait, as in scenario IV ($1,821,044), does not considerably affect the present value of net benefits compared to scenario II ($1,533,175), where the producer is taking advantage only of the Bt trait. The small difference in the present value of net benefits in favor of scenario IV is probably due to savings from reduced manual weeding. However, such savings are generally limited, for two reasons. First, the cost of increased herbicide use tends to offset the cost reduction of fewer manual weedings. Second, the savings from fewer manual weedings displayed in scenario IV are dependent on the availability of the glyphosate herbicide. If farmers do not have access to glyphosate, then they will probably revert to manual weeding, and the results will be similar for those in scenario II. Moreover, if we assume a higher technology fee in scenario IV because it takes into account both traits, then this scenario's present value of net benefits might actually be lower than that in scenario II.

Scenarios II and IV open for discussion whether a high enough technology fee may limit both the ability of farmers to capture technical gains (such as insect resistance or glyphosate tolerance) and the ability of the GM technology to compete with conventional cotton. This raises the policy question of whether the government of Uganda should intervene in the cotton market. The government may explore whether it is feasible to use targeted subsidies for the more resource-poor farmers to enable them to take advantage of the technology. As mentioned before, the same effect may be achieved by lowering the seed price to ensure that the technology is economically attractive. Both alternatives help reduce the financial risk of production for farmers.

In scenario V, as in scenario III, the technology fee is reduced to one-third of that charged internationally, which yields a present value of net benefits of $7,473,349, which is higher than in scenario I. One may explain this result by the additional benefit of using herbicides instead of manual labor for weeding, coupled with the reduced technology fee. In summary, all these scenarios underline that the size of the benefits is highly dependent on the technology fee paid by farmers.

The present value of net benefits grows to $32,242,210 in scenario VI, which doubles (1) the total cotton area and (2) the adoption rate, while keeping the organic cotton area fixed at 40,000 hectares. The purpose of this simulation is to emphasize the impact of increasing total crop area and adoption. We need to treat scenario VI, in which both area and adoption double, as an upper boundary for the estimation of potential gains from the adoption of GM cotton in Uganda.

Table 6.4 also presents the results from the sum of changes in total nominal benefits over the course of the study's simulations with respect to the counterfactual of no technology adoption. Total nominal benefits (without discounting to account for the time value of money) track the results from all the estimations of annual net benefits by summing nominal annual net benefits. The highest total nominal benefits were recorded for scenario VI ($297,960,600); the lowest values were for scenario II ($15,989,645) and scenario IV ($18,626,829).

In most developing countries, significant attention is focused on the probability that outcomes will be negative; that is, attention is focused on the

FIGURE 6.3   Distribution of the net present value of total surplus across scenarios

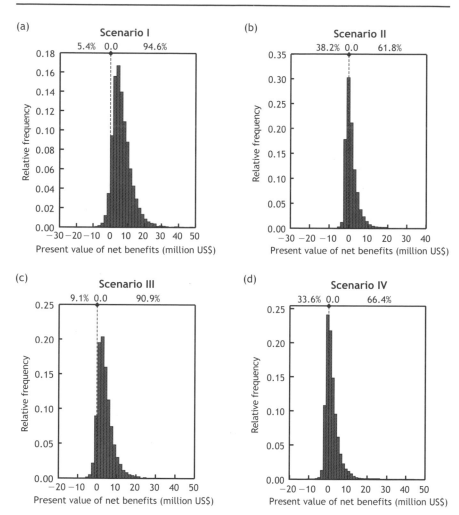

Source. Authors.

Notes: The vertical line with a diamond marker represents a zero present value of net benefits. Relative frequency refers to the probability for each bar. The sum of all probabilities in each histogram equals one. The percentages shown in the upper part of each graph represent the shares below and above the threshold of zero present value.

downside risk. Our study looked at the distribution of outcomes from the estimation of the present value of net benefits to examine the downside risk across scenarios (Figure 6.3). In scenario I, the probability of having a negative present value of net benefits is 5.4 percent. In the case of a public or private sector

FIGURE 6.3   (continued)

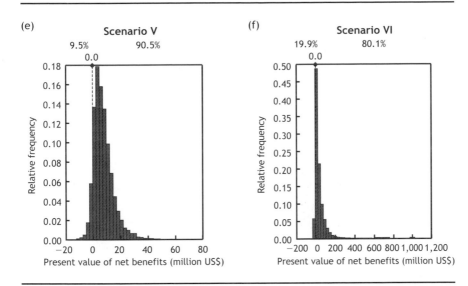

Notes: The vertical line with a diamond marker represents a zero present value of net benefits. Relative frequency refers to the probability for each bar. The sum of all probabilities in each histogram equals one. The percentages shown in the upper part of each graph represent the shares below and above the threshold of zero present value.

release where the developer charges a technology fee, as in scenario II, the probability of a negative present value of net benefits increases to 38.2 percent.

The latter is an interesting (yet trivial) result, because it indicates that charging a technology fee decreases the present value of net benefits compared to a no-cost technology release while increasing the probability of undesirable outcomes. However, it is important to maintain this scenario to show the potential gains to producers and to regulators of reducing technology fees when possible. In scenario III, the probability of a negative present value of net benefits is 9.1 percent. This probability is higher than in scenario I, but much lower than in scenario II. When one considers the situation of managing both the Bt and HT traits with a technology fee (scenario IV), the probability of a negative present value of net benefits is 33.6 percent. The downside risk in scenario V (9.5 percent) reinforces the idea that negotiating with the technology innovator for a lower technology fee increases the probability of having positive economic outcomes for society. Results from scenario VI show that the probability for a negative present value of net benefits is 19.9 percent. This is higher than in scenario V but lower than in scenario IV. Gains from increasing maximum potential area and adoption are counterbalanced by technology fee charges. The use of Bt/RR increases significantly the level of net benefits and

FIGURE 6.4   Distribution of internal rate of return

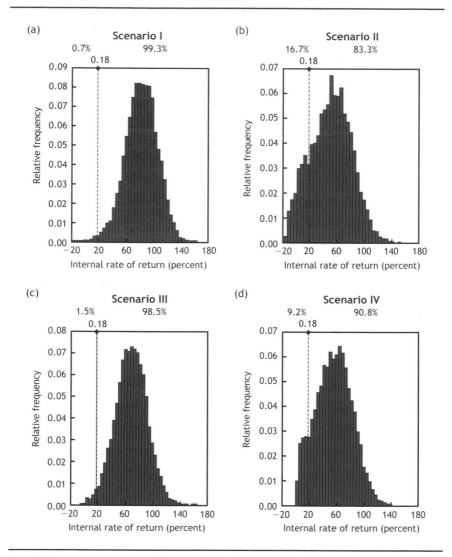

Source: Authors.

Notes: The vertical line with a diamond marker represents a threshold of 18 percent, which is the interest rate for loans set by the Central Bank of Uganda. The percentages shown above each graph represent shares below and above the threshold of 18 percent. Relative frequency refers to the probability for each bar. The sum of all probabilities in each histogram equals one.

the rate of return, but it also increases the level of downside risk. Further investigation is needed to examine this trade-off.

Results from the estimation of the IRR distribution mirror those for the distribution of the present value of net benefits. Scenarios VI, V, and I have

FIGURE 6.4 (continued)

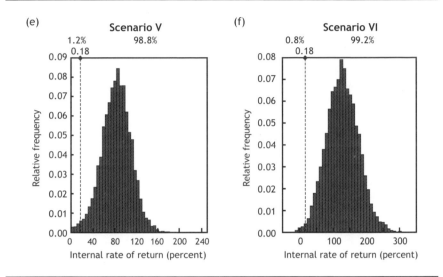

Source: Authors.
Notes: The vertical line with a diamond marker represents a threshold of 18 percent, which is the interest rate for loans set by the Central Bank of Uganda. The percentages shown above each graph represent shares below and above the threshold of 18 percent. Relative frequency refers to the probability for each bar. The sum of all probabilities in each histogram equals one.

the three highest average IRRs (131, 83, and 81 percent, respectively), whereas scenarios III, IV, and II have the three lowest IRRs (Figure 6.4). Note that in all scenarios, the peak estimated IRR is higher than the 18 percent interest rate for loans set by the Bank of Uganda (see Table 6.3). This rate represents the cutoff point for accepting or rejecting specific projects. Thus, an investor may reasonably pick any of these scenarios—usually the one with the highest IRR—in an investment decision process. Despite the IRR estimates being above the threshold interest rate, they are not as high as results obtained in other agricultural research studies found in the literature.

Figure 6.4 displays the distribution for the IRRs for all scenarios. For scenario I, there is a 0.7 percent probability that the IRR will be less than 18 percent, the interest rate for loans at the Bank of Uganda. In scenario II, the probability of having an IRR of less than 18 percent is 16.7 percent.

A rise in technology fees increases the probability that the scenario will not be funded. If the technology fee is reduced to one-third the level of those charged internationally, the probability that the IRR is less than 18 percent is 1.5 percent, as in scenario III. The 1.5 percent probability is higher than the public sector releases but significantly lower than scenario II with the full technology charge. Introducing herbicide management into the picture increases the probability to 9.2 percent (scenario IV). This scenario considers a technology fee that is the same as that charged internationally yet lower than the technology fee in scenario II. The benefit of herbicide management in this sense serves to dampen somewhat the impact of a higher technology fee.

If one introduces herbicide management and further reduces the technology fee to one-third of that charged internationally, the probability that the IRR will be less than 18 percent is 1.2 percent, as in scenario V. Finally, the probability that scenario VI will be less than 18 percent is 0.8 percent. Scenarios IV, V, and VI are linked to (and dependent on) the assumption that farmers will have ready access to the herbicide glyphosate. If access is limited and farmers are indeed charged a technology fee for the herbicide tolerance component, then one can expect the level of benefit to decrease and the riskiness of this scenario to increase compared to the results obtained in this simulation.

Figure 6.5 presents a sensitivity analysis for scenarios II and IV. The tornado graphs in Figure 6.5 express the relative impact of particular input parameters on the outputs from the simulations. The @Risk program regresses the present value of net benefits to each of the parameters with a probability distribution included in the simulation. The resulting parameter gives an indication of the relative strength of the relationship between parameters and outputs. Cotton yield, elasticity of supply, and cotton prices are the most important parameters: they explain changes in the present value of net benefits in both scenarios. Cotton yield and prices have a positive impact, whereas elasticity of supply has a negative effect on the present value of net benefits. A higher yield difference of the Bt/HT variety over a conventional counterpart has positive effects on the present value of net benefits, but the value of the coefficients decreases over time. This result is to be expected, because even though the study allows yield distribution to vary each year, the present value of the impact decreases as a result of the discounting process that values money less every additional year.

Figure 6.6 introduces the cumulative distribution functions for each of the datasets from the simulations corresponding to the six scenarios analyzed. As can be seen in Figure 6.6, no scenario dominates other scenarios using the first-degree stochastic dominance rule: all the cumulative distribution functions for the different scenarios cross one another.[13] Thus, one cannot provide an

---

13 Explanations of first- and second-degree stochastic dominance are provided in Appendix 6.

FIGURE 6.5   Analysis of present value of net benefits' sensitivity to significant distribution parameters included in simulations for scenarios II and IV

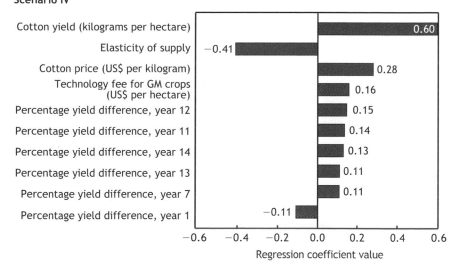

(a)
Scenario II

(b)
Scenario IV

Source: Authors.

Notes: We introduced the feature of having a separate and independent yield and cost difference distribution for each year in all six scenarios/simulations. The graph calculated by @Risk presented here considers only those distributions which are significant in explaining outcomes, so cost differences and several years' yield differences are omitted. The graph reports several percentage differences in yield between the GM cotton and a conventional variety that correspond to multiple years starting with year 1 of the adoption process in the simulation. GM cotton produces higher yields than conventional cotton, but the percentage yield difference declines over time in its impact on the present value of net benefits. This decline is a reflection of the discounting process, where yield differences in the future may count less than more recent ones.

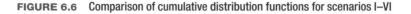

FIGURE 6.6   Comparison of cumulative distribution functions for scenarios I–VI

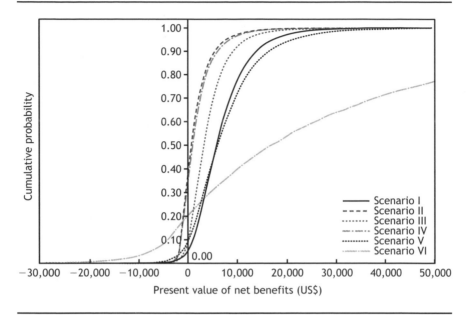

Source: Authors.

ordering of the different scenarios with this rule. The implication of the in-
conclusive results from the first-degree stochastic dominance rule is that one
needs to examine the simulations using the second-degree stochastic dominance rule to define an efficient set.

The program SIMETAR (© 2008) was used to examine the pairwise
comparisons between the different scenario simulations using this second-
degree rule. Table 6.5 describes which scenario, measured by this rule, dominates other scenarios. As can be seen from this table, most of the scenarios
dominate at least one other scenario, with the exception of scenario II, which
does not dominate any other scenario. In this sense, one can qualify scenario
II (public or private release with full technology fee charges) as not desirable
compared to the other scenarios. The explanation for this outcome is relatively
simple. A technology fee priced at an international level is significantly higher
than what a producer in Uganda may be able to pay, considering the low levels
of cotton productivity in the country. Scenario IV, in turn, dominates only scenario II, implying that unless producers are able to take advantage of both the
insecticide resistance and herbicide tolerance, a GM cotton variety that incorporates both Bt and HT traits may not be useful to producers in Uganda. If
producers are to take advantage of herbicide tolerance, complementary inputs

TABLE 6.5   Second-degree stochastic dominance for scenarios I–VI

Scenario	Scenario I	Scenario II	Scenario III	Scenario IV	Scenario V	Scenario VI
Scenario I		X	X	X		
Scenario II						
Scenario III		X		X		
Scenario IV		X				
Scenario V	X	X	X	X		
Scenario VI	X	X	X	X	X	

Source: Authors.

Notes: An X indicates that the scenario in the row head dominates the scenario in the column head. See Table 6.4 for a summary description of the various scenarios.

must be available, specifically the herbicide glyphosate. In turn, scenario VI dominates to a second degree scenarios I–V. This result implies that a decision-maker will prefer scenario VI over all other scenarios.[14]

The stochastic dominance analysis implies that farmers are better off having a free technology (scenario I) or a reduced technology fee (as in scenario III). Even more preferable is if farmers are able to take advantage of the cumulative benefits from insect resistance and herbicide tolerance while paying a reduced technology fee (scenarios V and VI). These results are not unexpected and conform to common sense.

Reducing the technology fee compared to the international levels charged to farmers is an appealing policy, although the issue is whether it can be done through legislation, as was done by the government of India, or whether it must be the product of market integration. Reducing the fee through market integration could be promoted by the ability to negotiate seed prices in a specific sector—negotiating capacity might have this effect in West Africa—or by other market forces operating in a country. The challenge, in any case, is ensuring that farmers benefit from the technology, if it is valuable, while also allowing seed companies to benefit sufficiently from technology sales to ensure the availability of a stream of technologies over time.

Figure 6.7 introduces an advanced sensitivity analysis of the present value of net benefits to 14 stepwise changes from the original values of the probability distribution of seed cotton prices and technology fees for scenario VI.

---

14 As described in the methods discussion, if the need arises to rank with more precision scenarios I–VI using second-degree stochastic dominance, the analyst needs to posit a range of reasonable risk-aversion coefficients. SIMETAR (© 2008) uses the stochastic efficiency with respect to a function approach to determine the range of risk-aversion coefficients and an assumed utility function that fits the data to allow choosing among alternatives.

FIGURE 6.7   Advanced sensitivity analysis of the present value of net benefits to changes in cotton lint prices and technology fees for scenario VI

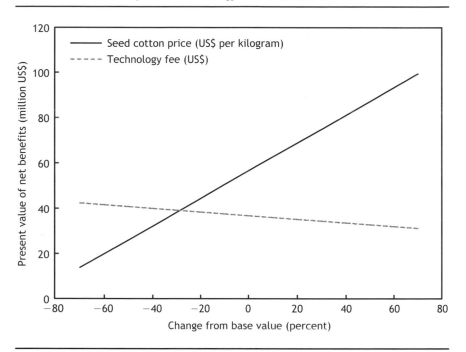

Source: Authors.

We conducted similar exercises for the other scenarios in the analysis, which are available on request. The advanced sensitivity analysis considered 14 stepwise departures representing changes to the base value of ±10, 20, 30, 40, 50, 60, and 70 percent for both seed cotton prices and the technology fee. For the base value and each of the fourteen stepwise departures from it, @Risk runs 10,000 iterations, for a total of 150,000 data points. Because we are analyzing two separate variables (cotton price and technology fee), the total number of data points is 300,000.

Results from the analysis in Figure 6.7 show that reductions of up to 70 percent of seed cotton prices decrease the present value of net benefits. However, in all cases, the present value of net benefits is still positive. In turn, increases of up to 70 percent in the technology fee caused a decrease in the present value of net benefits, but such values also remain positive.

Conclusions and Policy Recommendations

Results from simulations show that farmers in Uganda can gain from the introduction of GM cotton. Potential gains for farmers from such an introduction, reflected in positive simulation outcomes, depend on two critical parameters: current seed cotton yield in Uganda and the technology fee charged to farmers.

A critical need is for Uganda to improve current cotton yields and the overall productivity of its producers. The simulations showed, albeit indirectly, that it is hard for any innovation—be it conventional or GM cotton—to overcome low yields and thus generate gains for farmers in terms of finances or food security. Adoption of Bt or HT GM cotton can reduce insect and weed damage and ensure the enhanced production made possible by other productive inputs, such as fertilizers and soil preparation. One public policy goal is for the government of Uganda to establish programs—such as facilitating the use of fertilizer and biocides, or introducing practices that improve soil fertility—that result in the improvement of overall seed cotton yields by targeting input deficiencies.

As seen in the household survey conducted in Uganda, availability of herbicides and glyphosate is severely limited in the sample region of the study. This is perhaps unique to Uganda. It would be worthwhile to explore in more detail whether a program to improve access to herbicides is warranted. Another issue related to herbicide use that demands more detailed study is the substitution of herbicides for manual labor, as this would have social impacts. It would be necessary to examine who does the weeding (temporary workers, women, or children) and contrast the current situation of overall national labor surplus with localized labor shortages.

In the case of technology fees, the simulations strongly show that lower technology fees (compared to those charged internationally) indeed benefit farmers, not only by increasing the level of benefits but also by reducing the risk of negative outcomes or losses. For this reason, a negotiation process has to occur where developers and users in the country settle on the price of the technology fees that will be beneficial to all stakeholders. This negotiation should take into account information on production, productivity, insect-control costs, potential for damage to the crop, and other production parameters to establish a fee price that yields a fair outcome to all participants in the innovation, technology deployment, and adoption processes.

This study has used two distinct methods for the assessment of sector-wide impacts from the adoption of Bt cotton in Uganda. The first is stochastic economic surplus. This method is relatively parsimonious in terms of data, but it can address parameter uncertainty and production and financial risk in the sector. The expanded economic surplus method allows modeling of important effects of GM crop use, such as yield damage abatement. The main disadvantage—as applied—is that this model does not consider irreversibility and thus may need to be complemented by simulations using the real-options model. Additional disadvantages include the model's dependence on the availability of parameter data—or at least expert opinions on the potential range for such data—and the need to explicitly model producer heterogeneity in the simulation process.

The second assessment method is that of stochastic dominance. This is a very useful tool in terms of assessing decisionmaking under risk. The method is relatively easy to apply, provided there is access to computer programs to perform the analysis, and is very powerful in terms of discriminating between pairwise comparisons of policy options. If first-degree stochastic dominance cannot discriminate between policy options (or scenarios), analysts have to elicit a range of risk-aversion coefficients under which one policy option may be better than its pairwise comparison. Alternatively, this process can be approached by estimations using a stochastic efficiency function, as done in this chapter. The capacity to use both approaches is handy, especially when addressing the issue of downside risk for risk-averse producers.

Chapter 7

Alternatives for Coexistence of GM and Organic Cotton Production in Uganda

Guillaume Gruère

C otton is Uganda's third-largest export crop after coffee and tea (Baffes 2009). Although exports have fluctuated over time (see Figure 7.1), about 87 percent of the cotton lint produced in Uganda between 1996 and 2007 was exported. Despite low volumes of exports compared to other countries, the high proportion of cotton exported demonstrates the importance of trade in all cotton matters. In this context, it is legitimate to ask whether the introduction of genetically modified (GM) cotton would have any effect on Uganda's cotton exports. This chapter provides a brief analysis of this question, using available data, results from the literature, and the study's own assessment of the challenges ahead.

Trade Considerations: Would GM Cotton Affect Trade?

Conventional Cotton

First, it is important to note that there is no differentiation of GM and conventional cotton on the international cotton market (ICAC 2010). GM product regulations focus on novel food and feed items or on GM living organisms, therefore excluding nonfood processed items. Conventional and GM cotton have been commingled and sold together from the beginning. Cotton lint is priced based on quality and market supply and demand, not on whether it was derived from GM cottonseed.

Therefore, no reason exists to be concerned with export losses of cotton in general, either for Uganda or other countries, if GM cotton is introduced.[1] Even though most exports of cotton are purchased by buyers in Europe

---

1 As noted in ICAC (2010, 7):

> There are no price differentials for biotech and non-biotech cotton fiber or textiles containing biotech cotton. There is no evidence of rejection or price discrimination against biotech cotton by any segment of the market or region. In practice, markets do not identify biotech cotton content, but rather evaluate cotton properties based on quality characteristics.

FIGURE 7.1   Cotton production and trade, 1996/97–2007/08

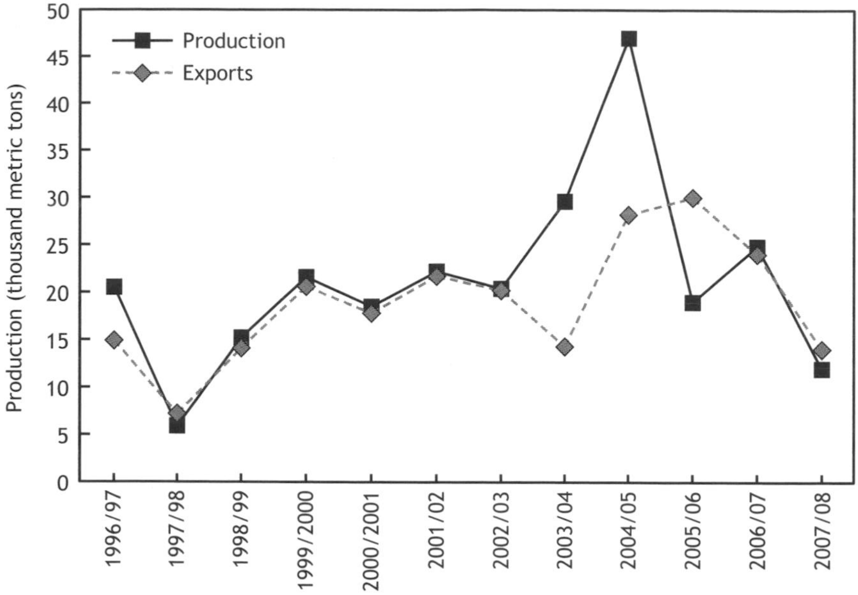

Sources: Baffes (2009); ICAC (2008).

(mostly the United Kingdom and Switzerland, as shown in Figure 7.2), the use of GM cotton would not result in any regulatory barrier for cotton lint.[2] The situation is different for cottonseeds, which are more closely regulated as "living modified [organisms]" under the Cartagena Protocol on Biosafety. Cottonseed oil derived from GM seeds is also subject to marketing regulations in Europe, including import approval, labeling, and traceability requirements. However, because Uganda does not export significant amounts of cottonseed or cottonseed oil to Europe, it is not expected to lose exports of cotton or cotton products in the near future.

In contrast, Uganda may lose on the world market if it does not adopt GM cotton. Other competing countries' adoption of GM cotton has contributed to an increase in their production and export, resulting in a deflation in world market prices (Frisvold, Reeves, and Tronstad 2006). For instance, the rapid adoption of GM cotton varieties between 2002 and 2008 has contributed to

---

2 Private standards may restrict access for cotton that may be GM to specific European buyers, but these are generally regrouped under fair trade and organic cotton standards (see next subsection).

FIGURE 7.2   Cotton exports by destination (major importers)

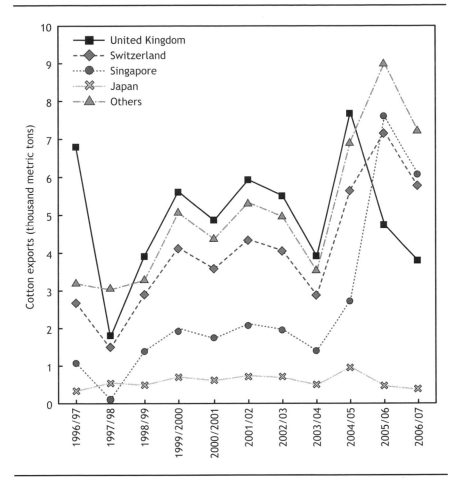

Source: ICAC (2008).

the doubling of India's average cotton yield, which resulted in India becoming the second-largest cotton exporter after the United States (Gruère, Bouét, and Mevel 2007). Using a simple three-country partial equilibrium trade model, Frisvold, Reeves, and Tronstad (2006) find that insect-resistant (Bt) cotton introduction in China and the United States resulted in a relative decrease in world prices. This phenomenon is bound to have made small exporting countries like Uganda lose because of their decreased market share and prices.

This competitive pressure is represented in a simplified comparative statistics partial equilibrium model of trade in Figure 7.3. Figure 7.3(i) represents the domestic market for a large exporting country (denoted by subscript a) that has adopted the technology (for example, India or the United States). Figure 7.3(iii) represents the domestic market for a small exporting country (for example, Uganda, denoted by subscript b); Figure 7.3(ii) represents the international market (denoted by subscript x). When the large exporting country adopts the technology, the resulting improvement in the country's productivity causes its domestic supply curve to shift outward from S_a to S'_a. Because country a is a large exporter, this adoption directly affects the world market, shifting the excess supply curve from ES to ES' (Figure 7.3[ii]). As a result of this shift, assuming no change in demand, world price declines from p_0 to p_1. Consequently, producers in the exporting country will gain as long as their cost reduction exceeds the price decline they face. Consumers in all countries gain from this price decrease, and producers in all other competing countries lose. In particular, producers in the nonadopting country with small exports (country b) lose export sales because of the price reduction. Their Marshallian welfare loss is represented by the trapezoidal area enclosed between the two world price lines, the supply S_b, and the vertical axis in Figure 7.3(iii).

In contrast, if country b decides to adopt the technology, its supply curve will also shift outward, from S_b to S''_b. In this case, the technology change will not affect the world price significantly, because the new adopter is a small exporter, but it will allow country b producers to compensate for the decline in prices by an increased productivity. They will gain if the technology allows for a cost reduction that overcomes the price decline (from p_0 to p_1) they faced because of the adoption of the large exporting country.

The total opportunity cost of rejecting the technology for country b can be expressed as the difference between the welfare changes associated with its market equilibrium when adopting the technology (located at q''_b and p_1) and the welfare changes associated with the market equilibrium when it does not adopt it (located at q'_b and p_1) while others do. Obviously, each of these sets of welfare changes will vary according to the technology effects, the rates of adoption of the technology, and the domestic and international elasticities of supply and demand.

Some empirical economic studies have shown that the opportunity cost of not adopting GM cotton when competitors do can be significant for African countries south of the Sahara (Elbehri and MacDonald 2004; Anderson, Valenzuela, and Jackson 2008). But most of these global trade studies use economywide simulation models called computable general equilibrium

FIGURE 7.3   Opportunity cost of adoption for a small exporter

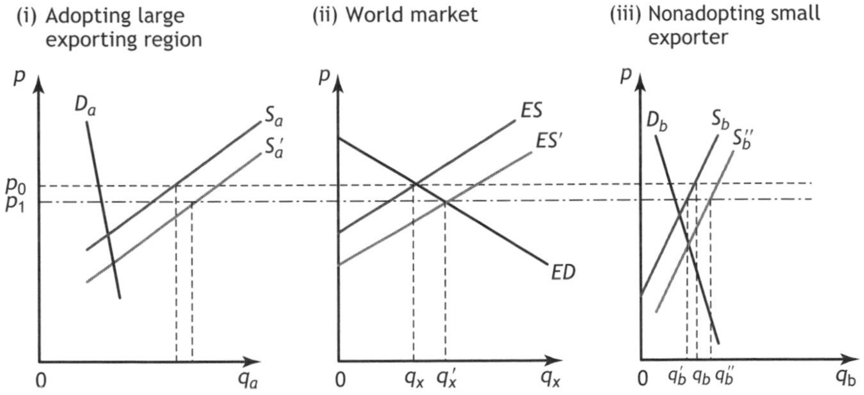

Source: Authors.

models (CGE) based on aggregate regional databases and thus do not provide insight as to what such costs would be for a specific country, such as Uganda.

Only two studies (Gruère, Bouët, and Mevel 2007; Bouët and Gruère 2011) provide simulation results for specific countries at a lower geographical level of analysis. Using multimarket CGE models, they measure the economy-wide effect of GM cotton adoption or nonadoption in Tanzania and Uganda when other countries adopt the technology. Table 7.1 provides a description of the assumptions and results of these studies.

In the first study, Gruère, Bouët, and Mevel (2007) focus on GM field crop adoption in Asia but also include Tanzania and Uganda as adopters of Bt cotton. Their simulation is based on the Global Trade Analysis Project (GTAP) trade database of 2001 and adoption rates in adopting countries as of 2005. However, they group the adoption of GM cotton with that of GM maize and soybeans.

Assuming a 30 percent adoption rate of Bt cotton in these two countries, with specific productivity assumptions (see the column on productivity shock in Table 7.1), and increased adoption of GM cotton, maize, and soybeans in other, competing countries, Gruère, Bouët, and Mevel (2007) find that adopting Bt cotton would lead to economic gains of about $44 million per year compared to nonadoption.[3] It is difficult to say whether this effect is driven

---

3 All dollars in this chapter are US dollars.

TABLE 7.1   Results of trade simulations of GM cotton adoption for Uganda and Tanzania

Study	Scenario in the study	Assumed adopting countries[a]	Productivity shock in Tanzania and Uganda (percent)	Welfare results for Tanzania and Uganda (million US$ per year)
Gruère, Bouët, and Mevel (2007)	Set A Scenario 1	Argentina (20), Australia (40), China (70), India (15), Mexico (61), South Africa (79), United States (81), plus adoption of GM maize and soybeans in many countries	None	+$0.46
	Set B Scenario 1	Same scenario, with Tanzania-Uganda (30), increased adoption in China (90) and India (25), and adoption of GM maize and soybeans in many countries	Yield: +15.5 Pesticide use: −23 Labor: −5	+$44.6
Bouët and Gruère (2011)	Scenario 1	Argentina (25), Australia (90), Brazil (40), China (75), Colombia (50), India (70), Mexico (64), South Africa (90), United States (93)	None	−$2
	Scenario 2	Same scenario, with Tanzania-Uganda (50) and five countries in West Africa (50)	Yield: +20 Pesticide use: −66 Labor: −10	+$5

Source: Authors.
[a]The percentage adoption rate for GM cotton in the country is given in parentheses.

only by cotton adoption. But assuming that the adoption of other GM crops in other countries does not have a strong effect, the total opportunity cost of not adopting Bt cotton is estimated to be about $44 million per year for the two countries. Naturally, because Tanzania is a much larger cotton producer, most of this gain would occur there and not in Uganda. For example, assuming that Uganda's production is only 20 percent that of Tanzania (based on averages of about 20,000 metric tons in Uganda and 100,000 metric tons in Tanzania), the opportunity cost of nonadoption of Bt cotton in Uganda would amount to $7.3 million per year.

In the second study, Bouët and Gruère (2011) focus on Bt cotton adoption or nonadoption in seven African countries south of the Sahara. Among the scenarios they consider are Bt cotton adoption by several countries, including Uganda and Tanzania (grouped as a region because of GTAP aggregation). They use a similar model to that in Gruère, Bouët, and Mevel (2007), with an updated GTAP database (from 2004) and updated adoption rates as of 2007/08 for all adopting countries. They assume a 50 percent adoption rate in Tanzania-Uganda, with a slightly higher productivity shock (see the productivity shock column in Table 7.1) than in Gruère, Bouët, and Mevel (2007). They find that Tanzania and Uganda would gain $7 million with adoption compared to nonadoption when the welfare results of $5 million from adoption are combined with avoiding the $2 million losses associated with nonadoption. Given the proportion of total production accounted for by cotton production in the two countries, they conclude that the opportunity cost of nonadoption of Bt cotton in Uganda is about $2 million. This estimate can be considered a more precise one than the estimate in Gruère, Bouët, and Mevel (2007).

Yet these much lower results with a higher shock and higher adoption rate may appear puzzling. These differences are likely due to the use of different reference years (there were changes in the cotton market between 2001 and 2005) and to differences in adoption outside Uganda and Tanzania. Indeed, Gruère, Bouët, and Mevel (2007) assume that India adopts at a rate of only 15–25 percent, whereas Bouët and Gruère (2011) use updated Indian adoption figures of 75 percent. Bouët and Gruère (2011) also include higher adoption rates in the United States, Brazil, and Australia, all major cotton producers. Higher adoption rates in these countries result in further depression of the price p_1, which results in lower welfare compensation for a small country (that is, country b in Figure 7.3). As a consequence, the surplus area of a shift in supply will be smaller if p_1 is lower. Thus, the longer a country waits for others to lead the trend to adopt, the more it loses and the less it will be compensated by adoption.

Overall, these differences in results are not large at the macro level, but they may imply significant welfare differences for individual cotton producers in countries like Uganda. Most importantly, they do demonstrate that Uganda would gain from adopting and would lose from not adopting. Bouët and Gruère (2011) do conduct a sensitivity analysis and find that changes in the productivity shock (or trade liberalization) do not affect the total welfare results in Uganda and Tanzania much, but that an increase in adoption from 50 to 75 percent would result in an additional $4 million net benefits in the two countries together.

Naturally, as in any CGE model simulation, all these results are based on a number of key assumptions, including (1) the use of known and stable productivity shock and adoption rates in each country of production, masking the variability among producers, production practices (input use), regions, and years and related uncertainties; (2) a perfectly competitive world market; and (3) a reference year of 2004 for the economic model. Thus, these results should not be taken literally and simply compared to the results of the household survey analysis. Their value is to provide some insight, based on specified assumptions, into the potential economywide effects of the use of Bt cotton in the region.

Organic Cotton

However, if conventional cotton exports will be unaffected, the introduction of GM cotton could potentially affect certified organic cotton production. As shown in Figure 7.4, the volume of organic cotton has been increasing from 2003 to 2008, but it still is relatively small compared to total production. The fast adoption of organic practices by farmers in 2006/07 (Gruère, Bouët, and Mevel 2007; CDO 2008) associated with the decline in total production has resulted in a steep increase in the organic share of the total production (reaching 20 percent in 2007/08, despite having remained less than 3 percent before 2006). The consequence of this rapid shift toward organic cotton, apart from reportedly having negative consequences on yields because of poor knowledge of organic techniques (CDO 2008), is that organic cotton has become a significant niche market for Ugandan

FIGURE 7.4   Organic and total cotton production, 2003/04–2007/08

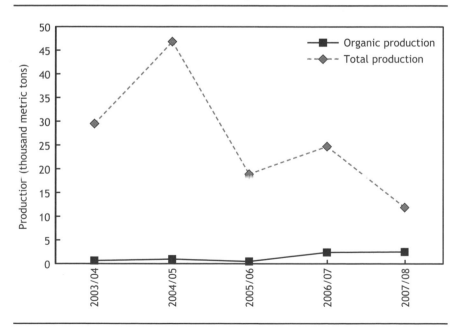

Source: Authors.

cotton producers. Whether this will still be the case in years to come is uncertain, but the presence of this niche poses the question of whether GM cotton introduction would be problematic.

Certified organic production requires the avoidance of GM seeds. Otherwise, organic cotton producers may be at risk of losing their price premium (assuming they do in fact receive a premium). So the relevant question is how the production could be organized to facilitate the coexistence of GM cotton and organic cotton in Uganda, so that producers can choose their production systems independent of each other. Other countries' experiences have shown that the two cotton production systems can successfully coexist—for example, India is the largest GM cotton adopter and the largest organic cotton producer—but there are still potential issues to consider to avoid costly commingling.

In-field coexistence may not be the largest challenge, but it can happen. As with other insect-pollinated crops, cross-pollination of cotton occurs, especially for fields in close proximity (less than 160 feet apart) (Ervin et al. 2010, 105). Care should therefore be taken to isolate the production of conventional non-GM seeds. Still, pollination may not be as significant a problem as human intervention, especially the mixing of seeds and the potential presence of GM cotton in a non-GM cotton field intended for seed production. The main issues really relate to the seed-marketing channel, where unintentional commingling can result in decertification. The next section discusses the issue of managing cottonseed marketing to ensure that certified organic cotton does not lose its premium in foreign markets.

Managing the Coexistence of Organic, Conventional, and GM Cottonseed Channels

Many papers have been published on the management of coexistence in the field between GM and non-GM crops (for example, see Demont et al. 2008). Most of these studies focused on food or feed crops (almost always GM maize) and the possibility of wind cross-pollination, in the context of the European Union. Only a few studies focus on developing countries (Falck-Zepeda 2006), and among those, scientific assessments of cross-pollinations are the most common. One study recognizes the role of gene flow stemming from human-driven seed dispersion and distribution in the case of GM maize (Dyer et al. 2009). Other studies analyze segregation requirements for grains postharvest, from the field to the processor and beyond (for example, Lin 2002; Huygen, Veeman, and Lehrol 2003; Wilson et al. 2008; Gruère and Sengupta 2009).

Yet all these studies focus on GM grains and oilseeds and do not really analyze the issue of coexistence in the cotton marketing chain.[4]

Unofficial seed flows of GM cotton seed have occurred in the past, especially in southern Asia, but virtually all reported cases were intentional movements driven by farmers' demand, not unintentional seed movements into non-GM markets. For instance, several unofficial Bt cotton varieties have been in circulation in western India (Pray et al. 2006). In fact, the first Bt cotton varieties planted in India were unofficial ones (Gruère, Mehta-Bhatt, and Sengupta 2008). Cotton growers in neighboring Pakistan used Bt cotton (copied from India) for some years before its official approval in 2009.

In the case of Uganda, with separate ginneries, organic lint is already separated from conventional, and introducing GM cotton does not present a specific new risk of mixing organic lint with other kinds. However, the issue of managing seeds (that may be recycled) is more challenging. After separation of cottonseed from cotton lint, organic and conventional seeds are pooled together. Part of the pool is used for oil crushing; part of it is sent to the Cotton Development Organisation (CDO) and then treated for recycling. The recycled seeds—which are not separated according to whether they are organic or conventional—go back to ginneries that deliver them to the farmers. The seed-marketing channel is shown in Figure 7.5.

If GM cotton is introduced, organic producers will not want their seeds mixed with GM seeds, whereas conventional cotton producers will see no need to separate their seeds from GM cotton seeds. The biotech companies could take either position on separating seeds. They might not want to have free recycling of GM cotton seed mixed with non-GM conventional (or organic) seeds, because such mixing could threaten their control of intellectual property rights. But if biotech companies find an alternative system to ensure these rights (such as a farmer/cooperative contract), they might not be as concerned about seed mixing (an advantage of recycling seed is that it eliminates the requirement to purchase new seed every year). Given this variety of possible views on seed mixing, three broad approaches are pos-

---

4 One report includes an analysis of the implications of introducing Bt cotton into the marketing channels of five countries that belong to the West Africa Economic and Monetary Union (Gruère and Cartel 2007). The issue raised by non-GM standards associated with organic and fair trade cotton is outlined. However, because the vertically integrated structure of the marketing chains in these francophone countries is completely different from the cotton chains in East Africa, their lessons do not apply to Uganda.

FIGURE 7.5   Original market chain for cotton seed before GM cotton introduction

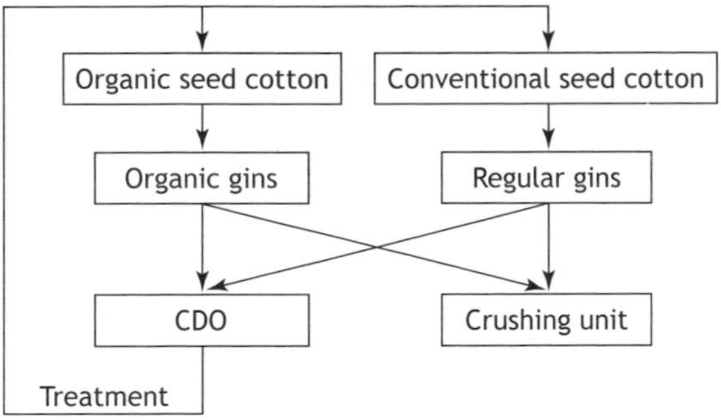

Source: Authors.
Note: CDO = Cotton Development Organisation; GM = genetically modified.

sible. First, organic seed could be kept separate from mixed conventional and GM seed to meet organic producers' need to avoid GM seed. Second, GM seed could be kept separate from both organic and conventional seed to meet both organic producers' need and the concerns of biotech companies about intellectual property rights. Third, both organic and GM seed channels could be separated from conventional ones. The three options are considered in what follows.[5]

Option A: Separating Organic Cottonseed

The advantage of this alternative is that the control focuses only on organic cotton, which is already separated from other cotton varieties at the output level (Figure 7.6). To make sure seeds do not include GM seeds, the process should separate organic cottonseed at the source: the ginnery level. Instead of pooling organic seeds with others, they would be kept separate from this stage. It would be preferential to just keep them at the ginnery, treat them, and then redistribute them to the organic producers. If this is done properly, assuming seed treatment can be done separately, organic producers should be able to obtain only non-GM seeds. However, if seed mixing might occur along the treatment chain, basic testing of seeds could be conducted on a sample before the seeds reach the organic farmers.

---

5 As a caveat, marketing cost differences (including transportation and other costs) are not explicitly taken into account, even if they would affect the design of an optimal system.

FIGURE 7.6   Market chain for cotton seed with GM cotton introduction:
Separating organic seeds

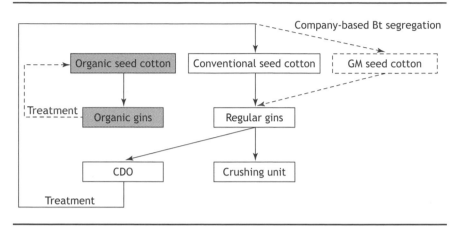

Source: Authors.

Notes: Shaded areas indicate separated sections of the market chain. Dotted lines indicate changes compared to the original market chain. Bt = insect resistant; CDO = Cotton Development Organisation; GM = genetically modified.

Option B: Separating GM Cottonseed

In this case, the process would focus on GM cottonseed (Figure 7.7). The advantage of this alternative would be that all other seeds would follow the same steps as before: only the new seeds would be affected. The point of control of GM seeds could be CDO, which is central to the seed recycling system. But if CDO has to separate GM seeds from organic and conventional seeds, it would require intensive testing at this stage with the potential for many errors.

A potentially more cost-effective alternative would be to have GM cottonseed separated at the ginnery level. This separation could be based on a basic procedure of physical separation rather than on systematic testing going from the field to the ginnery. This implies that GM farmers are able to transport their seed cotton from field to ginnery at the same time, which would be easier if these farmers were concentrated in the same area. The ginnery segregation could be done by differentiating the process in time. For instance, a certain period would be allocated to processing GM cotton and separating their seeds, and then the rest of the processing period would be focused on processing conventional cotton. At the end of the process, conventional seeds would be pooled together with those of organic cotton and then would be used for crushing or sent to CDO and treated for recycling. The GM seeds would be kept separated and potentially would be treated for

FIGURE 7.7   Market chain for seed cotton with GM cotton introduction: Separating GM cottonseeds

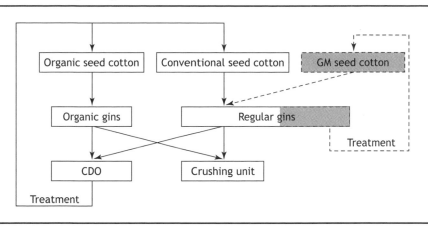

Source: Authors.

Notes: The shaded area indicates the separated section of the market chain. Dotted lines indicate changes compared to the original market chain. CDO = Cotton Development Organisation; GM = genetically modified.

recycling and sent to GM producers (assuming the biotech companies allow the recycling of GM seeds).

Option C: Separating Both Organic and GM Seeds

To sustain its rights over the seed, the developing company can propose an arrangement to separately handle the marketing chain of GM seed. This arrangement does not imply that there would be no risk of conventional or organic seed being mixed with GM seed. In the case of a separate GM marketing chain, both organic and GM seeds should be recycled or used separately from conventional ones, to preserve both organic producers' certification and the GM seed developer's rights. The problem with this double procedure would be the unnecessary separation of conventional seeds from all others for no particular purpose. A double separation would be greatly simplified if the production were composed only of organic and GM cotton.

What Option to Select?

The choice of option depends on (1) how the biotech company plans to sustain its rights over GM seed, and (2) the size of the market for organic versus GM seed. From an economic standpoint, assuming that the cost of segregation depends on the production volume, the least-produced type of cotton should be the one subject to segregation to ensure cost effectiveness. In principle, the onus should fall on either type but is likely to be on GM in the initial years after its introduction (with low adoption) and then on organic. This succession of systems would complicate the implementation.

In practice, the segregation costs usually fall on the pure product. For instance, the use of GM-food labeling regulations has resulted in non-GM segregation, because GM and conventional products would otherwise be mixed and sold together, and there is a specific demand for non-GM products that gives them an added value. But in this case, GM seed and organic seed can be considered pure products to a certain extent.

From a public economics perspective, the balance sheet is mixed. On the one hand, GM cotton is generating this new challenge, which can be seen as a production externality. On the other hand, the need for non-GM seed in organic production for a niche-market export could be seen as a consumption externality. The role of the state would be to make sure the best type of solution is achieved for both types of agents in the most cost-effective way. If the separation of organic seed is necessary but costly, whereas the separation of GM cotton from conventional is more difficult, option A should be taken, but there could be some type of compensation from GM production revenues to organic cotton production. If, however, the developing company runs a large risk of losing its rights, it will manage the segregation of its seeds (option B), leaving the organic and conventional sectors untouched.

Mismanagement of this issue threatens organic cotton producers with the possible loss of certification and therefore a loss of export revenues. It also threatens the biotech company with loss of the revenue from licensing of GM cotton. Because these risks are borne by both sides, they should be dealt with ahead of GM cotton commercialization, by bringing all the actors to the negotiation table. Establishing an arrangement—such as options A, B, or C proposed here—would ensure an optimization of marketing possibilities for the whole cotton sector.

Putting this debate in the context of Uganda, a country that does not use hybrid cotton varieties, it is highly likely that the developing company would take major steps to avoid losing rights to its seed and will follow option B. If this happens, the issue of organic cottonseed may resolve by itself. But as GM cotton adoption grows in the medium to long run, segregation at the ginneries may be more challenging and result in a situation where the risk of GM seed being mixed with conventional or organic seed will grow despite the developing company's efforts to protect its rights. The only way to address this situation will then be to also separate the organic seed-marketing channel (option C).

Conclusions and Policy Recommendations

This chapter has reviewed the possible trade considerations with the introduction of GM cotton in Uganda. As noted, there is no differentiation of GM and non-GM cotton lint on the international cotton market. Reviewing the trade and economic literature turns up evidence from two simulation studies showing that the opportunity cost of nonadoption of GM cotton could amount to between $2 and $7.3 million per year for Uganda.

However, this chapter has argued that the introduction of GM cotton could affect the market for organic cotton (which forbids the use of GM seed). The critical issue to manage is the production and especially the distribution and recycling of seeds. Because organic cotton lint is already separated, the main issue would be related to the separation of GM and organic seeds. Three alternative scenarios were considered: separation of organic seed, separation of GM seed, and separation of both organic and GM seeds.

Comparing these options, each of them may be optimal under certain conditions, as the best alternative will depend on the developing company's strategy for controlling its intellectual property rights and on the volumes of organic, conventional, and GM seed produced. That said, the Ugandan context suggests that a biotech company developing a GM cotton variety would by itself take measures to separate its seed to secure its rights, meaning that the seed-marketing channel would not need public intervention, at least in the short to medium run. At the same time, once GM cotton is adopted on a larger scale, organic seed would have to be separated to ensure that organic producers' certification remains valid.

The main recommendation is that, although there is no risk of export losses for cotton, policymakers should actively engage in a dialogue with both the company providing a GM cotton variety and organic producer associations, as well as other stakeholders (such as CDO), to discuss and anticipate the issue of managing seed marketing and coexistence before the introduction of GM cotton. This strategy will ensure that cotton producers can choose their production systems independently of each other and that organic cotton producers will not lose their access to export markets.

Chapter 8

Conclusions and Recommendations

Daniela Horna, Patricia Zambrano, and José Falck-Zepeda

The design of biosafety regulatory frameworks needs to incorporate a well-defined process for evaluating genetically modified organisms and to be flexible enough to allow the use of innovative approaches that can be adapted to the changing technology environment. Providing support to the biosafety regulatory process in developing countries with limited or no biosafety framework is of critical importance, as most of these countries are parties to the Cartagena Protocol on Biosafety and, as such, are required to develop their own biosafety assessment protocols. Countries may opt—and in fact some developing countries have already opted—to include socioeconomic assessments in their biosafety regulatory process. This monograph can support socioeconomic assessment practitioners, regulators, and policymakers in the approval process of genetically modified (GM) technologies. We use GM cotton in Uganda to illustrate the methods and results of a socioeconomic assessment. It was the National Biosafety Committee of Uganda that originally requested an assessment of the socioeconomic impact of adopting GM cotton. At the time, the committee was deliberating on the approval process of a GM cotton confined field trial. This monograph contributes to the regulatory process in two ways. First, it provides a good—although preliminary—understanding of the likely impact of growing GM cotton in Uganda. Second, the monograph outlines a methodological reference for implementing future socioeconomic assessments in the context of a biosafety decisionmaking process.

About the Introduction of GM Cotton in Uganda

The potential impact of GM cotton adoption in Uganda was done taking into account first the institutional environment and then the different sectors of society, including farmers' perspectives, the aggregate economy, and trade considerations. Primary and secondary data sources were used or consulted for the development of this monograph. Findings suggest that GM cotton has the potential of contributing to the improvement of cotton productivity in Uganda.

As discussed in this monograph, there are binding productivity constraints that may limit the full exploitation of the potential benefits that this particular technology can bring to cotton farmers in Uganda; limit the potential increases in productivity that may move farmers to an efficient productivity frontier, where they use inputs efficiently and have higher benefits than costs; or both. Certainly, GM cotton can be viewed as a risk management option that can be used to reduce downside risk for farmers in Uganda by reducing the potential losses stemming from weeds, the bollworm complex, and other target pests addressed by the specific traits inserted into cotton varieties.

Institutional Concerns about Delivery and Adoption of GM Technologies

Institutional constraints need to be addressed for biotechnology to have an impact (see Chapter 4). Although the technology comes from the private sector, successful diffusion of GM cotton depends sensitively on strong public institutions, including clear support from the state, a dynamic research sector, and provision of extension and credit services. As the Cotton Development Organisation (CDO) controls seed cotton distribution and cottonseed commercialization, it has a direct and major influence on the final adoption of GM cotton varieties by farmers. CDO's formal position on GM cotton is not entirely clear, though it appears to have a positive attitude toward the technology. At the same time, at least in the past, CDO has also supported organic cotton production.

To facilitate the successful adoption of the technology by farmers, the public institutions involved in the cotton chain, mainly CDO and the National Agricultural Research Organisation, need to have a plan of action for GM seed distribution. Three decisions are especially critical: (1) the variety to use for introducing the genetic trait, given CDO's one-seed policy (for example, would the *Bt* gene have to be introduced to the BP52 cotton variety, or would the country consider disregarding the one-seed policy and introducing other Bt [insect-resistant] and conventional cotton varieties already available in other countries?); (2) the level of support to provide to organic cotton production; and (3) the method of public or private delivery of the technology, which will influence the price of the seed. These decisions would have to be made in parallel with the progress of the regulatory process. Delaying these important decisions would most likely have consequences for the benefits and costs of the adoption of GM cotton and its potential success in Uganda.

The last of the three decisions, the delivery method, is particularly relevant because of its effect on the price of the seed or technology fee. Uganda needs to develop a fee negotiation strategy before the developer releases the technology to farmers. Although there is no clear indication of what the fee for Bt cotton will be, it is unlikely that the developer will release the technology for free. Estimates from stochastic budget analysis and the economic surplus model show that in addition to the yield level, the technology fee is a critical determinant of both the level of benefits to the society and the downside risk to producers (see Chapters 5 and 6).

On the regulatory side, the confined field trials for GM cotton were approved and implemented in 2009 after two years of negotiations. The Biosafety Bill is still in evaluation, and the process, not yet completed, has taken more than seven years so far. Although quick and dirty processes are not helpful, decisionmakers need to be aware that the costs of these lengthy regulatory lags detract from the benefits of the technology. As has been mentioned a couple of times in this monograph, the inclusion of socioeconomic considerations has the potential to further extend these lags.

Cotton Productivity and Profitability

Cotton is an agricultural commodity that has a long tradition in some regions in Uganda (see Chapter 3). Growing cotton was a very important economic activity up to the early 1970s, but currently, production and price variabilities have created uncertainty in the cotton sector, making conventional cotton production highly risky. Not surprisingly, conventional producers are moving to other crops or to nonagricultural activities. In contrast, organic cotton production, although quite small, has been expanding for the past 15 years, opening new market possibilities for farmers in marginal areas. The available literature seems to suggest that few other crops could adapt to the agroenvironmental conditions in areas where organic cotton has been successful.

In the selected survey sites, cotton production under either the traditional or organic system shows low profitability (see Chapter 5). The main reason for this low profitability is poor and variable yield performance: a significant proportion of farmers interviewed have relatively low yields. These poor yields are explained by the low use of productive inputs, namely fertilizers and insecticides.

GM cotton has the potential to increase yields. The adoption of Bt cotton may lead to higher yields due to decreased vulnerability to targeted insects. Bt cotton's potential contribution is limited, however. Although bollworm, the pest to which Bt cotton is resistant, is an important pest in Uganda, there are several other economically important insects. More important, a damage-abatement technology, such as GM cotton, will reduce pest damage but will probably not increase yields above those obtained in the absence of such pests. Significant productivity improvements require making production inputs like fertilizers and pesticides available to producers.

Herbicide-tolerant (HT) cotton can be a good alternative to conventional or organic cotton, as it releases labor from weeding, a back-breaking and time-consuming activity. Producers can reduce the costs from family labor, hired labor, or both and can be employed more efficiently in growing other crops or in other economic activities. The potential benefits from HT technologies depend on labor costs and the affordability of herbicides. The prohibitive expense of herbicides is demonstrated by farmers' lack of experience in herbicide use; this underscores the need for improving access to appropriate extension and credit services.

The underlying assumption in our simulations is that the GM variety is the same as the local one (Bukalasa Pedigree Alba) except for the incorporation of the insect-resistant trait. Only if Uganda discards the one-variety policy and allows other more productive varieties could farmers perceive the pure yield effect from other outperforming varieties. For a successful technology release, Uganda will need to implement ancillary programs that will help farmers access complementary inputs to support the release of the GM cotton variety, including access to rural credit (see Chapters 5 and 6).

Other production and postharvest constraints must also be addressed to improve the profitability of the crop. It is necessary, for instance, to improve the ginning quality in the country. Ginneries are now operating under full capacity, and the equipment used is rather old, which affects the ginning turnout. This is currently a limiting factor, given that seed obtained after ginning is often recycled and distributed to farmers. Even apart from the ginning practices, ensuring seed quality is a problem in Uganda, and whether quality standards might pose a constraint on GM cotton adoption remains to be seen.

Coexistence of GM Cotton and Organic Cotton

The introduction of GM cotton does not pose a risk to exports of conventional cotton, but it might affect the organic cotton exports (see Chapter 7). Coexistence of both systems is possible if institutional arrangements are implemented to avoid commingling of the material. There seem to be incentives for both industries, GM and organic, to keep their outputs separated. The GM technology owners would be interested in realizing their intellectual property rights over their materials, while the organic sector needs to keep their cotton GM free to maintain their certification to export organic crops to the world market. Their interest in keeping the certification, therefore, will probably be determined by the organic production volumes they can achieve. The current freedom of ginneries to switch from conventional to organic and back, as appears to be practiced by some, must come to an end if GM cotton is commercialized and begins to mix with conventional cotton.

Methodological Challenges and the Inclusion of Socioeconomic Considerations in a Biosafety Regulation

In Article 26.1, the Cartagena Protocol on Biosafety states that countries may take into account socioeconomic considerations in the biosafety assessment process. Many developing countries, including Uganda, have not yet determined whether and how to include these socioeconomic considerations in their regulations. Several questions remain unanswered as countries advance in the process of approving GM technologies: At what stage of the regulatory process should socioeconomic consideration be included? What is the right level of analysis and scope for these socioeconomic considerations? How are they going to enter into the decisionmaking process? How do they relate to the results of the biophysical evaluations (experimental plots)?

There are clear differences between socioeconomic (impact) evaluations that are considered as part of a regulatory process and evaluations done outside of this context for examining technology performance. Protocols for socioeconomic impact evaluations done in the context of a biosafety regulatory approval process have to consider the amount of time needed to perform the evaluation and the likely binding budget constraints to perform such assessments. At the same time, developing countries have binding budget, human resource, and time constraints on the implementation of socioeconomic assessments, above and beyond what a similar study in an industrialized country will likely encounter. As shown in the small but growing literature, there is an inverse relationship between regulatory intensity (and thus cost) and the level of technologies deployed for farmers' use. Furthermore, regulatory uncertainty when the assessment process is ill defined will likely reduce the number of technologies deployed by the private and public sectors and thus reduce those technologies available to farmers.

Not only carrying out biosafety assessments but also interpreting their results presents problems. When the socioeconomic, food/feed, and environmental risk assessments show poor results, the regulatory decision not to approve the technology for release to farmers might be an easy one. However, the regulatory decisionmaking process might be quite ambiguous when the results are mixed.

A different question that needs to be answered by the regulatory approval process is what happens if the socioeconomic assessment yields a negative outcome, but this outcome is predicated on the institutional setting, not on the technology itself? Would a decisionmaking process render a do-not-release decision? Or would this type of assessment be a call for governments to actually solve the underlying institutional constraints that are likely binding not only GM but conventional technology as well? Furthermore, who should resolve institutional constraints: developers, government, or farmers? These issues require even more research, discussion, and debate with stakeholders in Uganda and other developing countries.

Implementing Socioeconomic Assessments: Methods and Tools

From a disciplinary standpoint, the discussion here centers on how to choose appropriate methods and tools that can provide robust results. Again, although it is important to have an evaluation based on farm-level information that is scientifically robust—verifiable and defensible—it is also important to propose methods and instruments that will not add delays and therefore costs to the regulatory process. A goal in this monograph was to try to address this need.

Clearly, no single method can be used to provide a definitive answer about the impact of GM technologies. Rather, the methodological framework proposed in this monograph comprises several methods and interdependent layers of analysis. Each method proposed has limitations, but the combination of methods renders a robust result. Below are some major factors to take into account when implementing similar evaluations.

- Before doing the assessment, it is important to gather information and identify factors affecting adoption that normally are not included in ex ante evaluations. This information is used to elaborate the assumptions about adoption that the study relies on. We believe that evaluating how the institutional setting could affect the technology's deployment is a critical first step in developing a robust impact assessment framework because: (1) researchers in charge of the assessment can build the assumptions that the analytical models rely on (yield, costs, prices, and so forth) and (2) policy- and decisionmakers will have a preliminary idea of how to deploy the technology in a way that takes advantage of the institutional setting and ensures as much as possible that the technology will have the impact predicted by the analytical models.

- The farm-level analysis has a potential problem with statistical bias. This issue is more critical for ex post assessments, where the better-off farmers may be the first adopters. In the case of ex ante assessments, such as the one included in this monograph, the issue of bias is still relevant, especially when conducting a small survey. As in any other impact evaluation, sample size and selection influence the results and thus the final recommendations. The sample size used for this evaluation was relatively small, and it did not allow for extending the conclusions to the country level. The sample size was particularly small for the case of organic producers. The selection of sites for implementing the surveys is critical. It is important to identify study sites that can represent the agroecological conditions of the country.

- It is critical to know how to estimate or derive the damage-abatement capacity of the technology, which at the farm and industry levels is referred to as *technology efficiency*. In an ex ante evaluation, technology efficiency can be based only on assumptions and references from the literature, as the technology has not yet been adopted. The impact of technology efficiency on production is significant and thus an estimate of it needs to be as accurate as possible. Given the uncertainty of the parameters used in the ex ante estimation, we used distributions and stochastic analysis, as they are the best alternative to deal with the lack of information. These methods provide possible ranges within confidence intervals, rather than single-valued results.

- The use of stochastic partial analysis conveniently deals with the issue of temporal and spatial variability of farm conditions. Failure to address this issue is one of the main criticisms of nonstochastic partial analyses. The stochastic analysis, however, does not allow for hypothesis testing. The results of a stochastic analysis are ranges of values for the outcome of interest. These results do not have a significance level attached to them, but they may be subject to further analysis and compliance with a set of robustness criteria.

- At the aggregate industry level, the reliability of the findings of the economic surplus model is limited, because the accuracy of the results depends on the extent to which the underlying parameters represent local conditions.

Given the nature of ex ante studies, parameters are pulled from previous studies and have to be adapted from related research conducted in other countries, or they can be elicited through interviews with local experts.

This study highlights the challenges of addressing heterogeneity in farmers' conditions in ex ante evaluations, given limitations on the collection of information to address the diversity of conditions. The information needed to disaggregate impact by region and target pest is not readily available in Uganda. Some secondary information could be gleaned from the household surveys, but the limited scope of our survey (detailed in Chapter 2) constrains extrapolation to the country as a whole. We discuss how the variations in some farmers' conditions, either social or ecological, could affect the adoption of GM cotton, but our results are only a rough estimate of this technology's impacts. Including heterogeneity in the analysis would require a larger sample, far greater than the one available, and the collection of more detailed information. Also, the financial expenditure on this type of study becomes a political decision. The public sector has to decide how much to invest in the approval process of GM technologies. The question to be answered is: what is the critical level of information that the competent authority would be willing to accept to approve (or reject) the technology?

An important caveat for this study is that it has not covered other relevant socioeconomic aspects, such as the effect of GM adoption on biodiversity and public health and the impact of consumers' perceptions of GM food. The importance of gender in the adoption process is briefly examined. There is already literature on gender evaluation and consumer perceptions that can and should be adapted to this methodological framework. However, methods and tools for the assessment of effects on human health and diversity need to be further developed.

Another important caveat is that the real socioeconomic impact of GM adoption may be influenced by production and market factors that cannot be predicted ex ante. Researchers should go beyond ex ante assessment to conduct in-depth analysis during the early stages of commercializing transgenic cotton to ensure that conditions are in place for having the best outcome possible and to address potential bottlenecks in a timely fashion. This in-depth analysis should include production and also key institutional issues and their effect on the introduction of GM technologies.

Household Survey Instrument

Program for Biosafety Systems (PBS-East Africa)

Ex ante Assessment of Bt-Cotton in Uganda
Household Questionnaire

ENUMERATOR DETAILS

Enumerator name: _____ Code: _____

Household Code No: _____ Date: _____

HOUSEHOLD LOCATION

	Name	Code		Name	Code
Village (LC1)			District		
Parish			Region		
Sub-county					

See codebook for village, parish, sub-county, and district codes

Region codes: 3 = northern, 4 = western

GPS READING

Latitude: _____ Longitude: _____

Altitude: _____ Accuracy: _____

TEAM LEADER DETAILS

Name: _____ Signature: _____

Date: _____

DATA ENTRY DETAILS

Data enter: Name: _____ Code: _____

A   General Information

A.1   Household Identification (Establish the following information)

Variable		Codes
Demographic data		
A.1.1 Name of household head		
A.1.2 Sex of household head		1 = female, 0 = male
A.1.3 Age of household head		
A.1.4 Age of spouse of head[1]		
A.1.5a Education level of household head		1 = no formal education, 2 = some primary education, 3 = completed primary education, 4 = some vocational or teacher training, 5 = completed vocational or teacher training, 6 = some O-level secondary education, 7 = completed O-level secondary education, 8 = some A-level secondary education, 9 = completed A-level secondary education, 10 = post-A-level education
A.1.5b Education level of spouse of head		
A.1.5c Highest level of education attained by any household member		
A.1.6a Primary source of income of household head		See codebook
A.1.6b Primary source of income of spouse of head		See codebook
A.1.6c Primary source of income for the household		See codebook
A.1.7a Household size		A household includes all members of a common decisionmaking unit (usually within one residence) that are sharing income and other resources. Include workers or servants as members of the household only if resident at least six months in the household.
A.1.7b Number of males aged 16 years and above		
A.1.7c Number of females aged 16 years and above		
A.1.7d Number of members aged below 16 years		
Survey respondent		
A.1.8 Name of primary respondent		
A.1.9 Sex of the respondent		1 = Female, 0 = Male
A.1.10 Age of the respondent		
A.1.11 Relationship to household head		1 = head, 2 = spouse, 3 = father, 4 = mother, 5 = son, 6 = daughter, 8 = in-laws, 9 = sister/brother, 10 = grandchildren, 11 = servant, 12 = other (specify)

[1] In case of more than one spouse, footnote details of other spouse

B Number and Value of Crop/Cotton Enterprise Equipment

	Rented		Owned			
	Number	Average rental value (USh)	Number	Purchase price (USh)	Age (Years)	Average maintenance cost per season (USh)
Agricultural enterprise equipment						
B.1 Hoes						
B.2 Ox-ploughs						
B.3 Tractor, including tractor plough						
B.4 Pangas						
B.5 Wheelbarrows						
B.6 Knapsack sprayer						
B.7 Watering can						
B.8 Irrigation system						
B.9 Rakes						
B.10 Slashers						
B.11 Gumboots						
B.12 Transport equipment for agricultural enterprise Specify:						
B.13 Other agricultural equipment						

For group assets record share owned by the household.

H   Access to Infrastructure and Services

Please indicate the distance from your residence to the nearest infrastructure and services below.

	a. Distance (km)[1]	b. Quality 1 = very good 2 = good 3 = average 4 = bad 5 = very bad
H.1.1 All weather road		
H.1.2 Seasonal road		
H.1.3 Bank or MFI		
H.1.4 SACCO[2]		
H.1.5 Government extension / agriculture / livestock office		
H.1.6 Input supply shop		
H.1.7 Ginnery		

[1] 1 mile = 1.6 km
[2] SACCO is Savings and credit cooperatives

C.2 Indicate the type of cotton production practices/technology household is aware of and is or is not yet using for cotton production

To enumerator: First establish whether practice/technology from the checklist on page 6 is mentioned or not. Only list what has been mentioned.

What are the household's current cotton production practices?	PRACTICE CODE	C.2.0 Source of advice/ info / tech2	C.2.1 Year first became aware*	C.2.2 Ever used it? 1 = Yes, 0 = No If no, go to C.2.4	C.2.3 If yes, year of first time use	C.2.4 If no, reason for not using	C.2.5 Used now? 1 = Yes, 0 = No If no, go to C.2.7	C.2.6 If Yes to C.2.5, average intensity of use (acres, share marketed, % processed, etc.)			C.2.7 If No to C.2.5, reason for stopping using practice	Code
								a. amount	b. unit	c. unit code		
Cotton management practices												
Cotton protection practices												

C.2.0 (source codes): 1 = NAADS service providers, 2 = Government extension workers, 3 = Farmer group members, 4 = NGO (specify), 5 = Other farmers, 6 = Radio, 7 = Demonstration sites, 8 = Technology development sites, 99 = Other (specify)

*Indigenous knowledge = 88

C.2   Indicate the type of cotton production practices/technology household is aware of and is or is not yet using for cotton production (continued)

What are the household's current cotton production practices?	PRACTICE CODE	C.2.0 Source of advice/ info / tech?	C.2.1 Year first became aware*	C.2.2 Ever used it? 1 = Yes, 0 = No If no, go to C.2.4	C.2.3 If yes, year of first time use	C.2.4 If no, reason for not using	C.2.5 Used now? 1 = Yes, 0 = No If no, go to C.2.7	C.2.6 If Yes to C.2.5, average intensity of use (acres, share marketed, % processed, etc.) a. amount	b. unit	c. unit code	C.2.7 If No to C.2.5, reason for stopping using practice	Code
Soil fertility management												
Soil and water conservation												
Post harvest handling												
Marketing and agroprocessing												

C.2.0 (source codes): 1 = NAADS service providers, 2 = Government extension workers, 3 = Farmer group members, 4 = NGO (specify), 5 = Other farmers, 6 = Radio, 7 = Demonstration sites, 8 = Technology development sites, 99 = Other (specify) ———

*Indigenous knowledge = 88

D   Cotton Production on Household's Farmland (Includes Land Rented-In or Borrowed)

D.1   Details on parcels of land cultivated by the household (annual and perennials): list all parcels the household cultivated (owned or rented-in) in the LAST 2 COTTON SEASONS

Note: A parcel is a contiguous piece of land operated by the household

Parcel name	Parcel ID#	D.1.1 Total parcel area (acres)	D.1.2 Distance from home (km)*	D.1.3a Planted to cotton in 2007 season? Yes = 1 No = 2	D.1.3b Planted to cotton in 2006 season? Yes = 1 No = 2	D.1.4a Irrigated in 2007 season? Yes = 1 No = 2	D.1.4b Irrigated in 2006 season? Yes = 1 No = 2	D.1.5 Tenancy: 1 = own with title 2 = own without title 3 = borrowed 4 = rented 9 = other (specify)	D.1.6 Tenure type: 1 = customary 2 = freehold 3 = mailo 4 = leasehold 9 = other (specify)	D.1.7 Land value if D.1.5 = 1 or 2, or rental cost if D.1.5 = 4 (USh)

*1 mile = 1.6 km

D.2 For the COTTON SEASON OF 2007, list all the plots allocated to cotton for all parcels cultivated by the household. Please also list all crops intercropped with cotton on each plot

Note: A plot is a contiguous piece of land operated by the household and has the same cropping system. Hence a parcel may have more than one plot allocated to cotton if the cropping system is not the same or if the cotton plots are not adjoined.

Parcel name	Parcel ID#	Plot ID#	D.2.1 Share of plot in parcel (%)	D.2.2 Cropping system[1]	D.2.3 Cropping system (use -9 if not applicable) and major intercrop	Intercrop code	D.2.4 % of plot area under cotton	D.2.5 Use recommended practices: row planting / spacing? 1 = Yes 0 = No	D.2.6 Source of seed / planting material[2]	D.2.7 Distance from residence to source of seed (km)	D.2.8 What costs did you incur in reaching source? (USh)	Seed / planting material D.2.9 Quantity (kg)	Seed / planting material D.2.10 Total value (USh)	Production/output D.2.11 Quantity	Production/output D.2.12 Unit	Production/output UNIT CODE	Production/output D.2.13 Unit conversion to kg	Production/output D.2.14 Total value of output (USh)	Plot control D.2.15 Who controls the plot? 1 = Male 0 = Female	Plot control D.2.16 R-ship to household head

[1]Cropping system: 1 = Pure stand (mono cropping); 2 = Intercropping (two crops); 3 = Mixed cropping (more than two crops); 4 = other (specify) _____

[2]Source of planting material: 1 = Bought; 2 = Saved from own harvest; 3 = Given by NGO; 4 = Given by NAADS/NGO working for NAADS / Private NAADS service provider; 5 = Given by government; 6 = Given by a friend/relative; 7 = Ginnery; 8 = Cooperative association, specify _____ ; 9 = Seed dealer; 10 = Trader not specializing in seed; 99 = Other (specify) _____

D.3 For the COTTON SEASON OF 2006, list all the plots allocated to cotton for all parcels cultivated by the household. Please also list all crops intercropped with cotton on each plot

Note: A plot is a contiguous piece of land operated by the household and has the same cropping system.
Hence a parcel may have more than one plot allocated to cotton if the cropping system is not the same or if the cotton plots are not adjoined.

Section	Column	Data
Plot control	D.3.16 R-ship to household head	
Plot control	D.3.15 Who controls the plot? 1 = Male 0 = Female	
Production/output	D.3.14 Total value of output (USh)	
Production/output	D.3.13 Unit conversion to kg	
Production/output	UNIT CODE	
Production/output	D.3.12 Unit	
Production/output	D.3.11 Quantity	
Seed/planting material	D.3.10 Total value (USh)	
Seed/planting material	D.3.9 Quantity (kg)	
	D.3.8 What costs did you incur in reaching source? (USh)	
	D.3.7 Distance from residence to source of seed (km)	
	D.3.6 Source of seed/planting material[2]	
	D.3.5 Use recommended practices: row planting / spacing? 1 = Yes 0 = No	
	D.3.4 % of plot area under cotton	
	Intercrop code	
	D.3.3 Cropping system (use -9 if not applicable) and major intercrop	
	D.3.2 Cropping system[1]	
	D.3.1 Share of plot in parcel (%)	
	Plot ID#	
	Parcel ID#	
	Parcel name	

[1]Cropping system: 1 = Pure stand (mono cropping); 2 = Intercropping (two crops); 3 = Mixed cropping (more than two crops); 4 = other (specify)

[2]Source of planting material: 1 = Bought; 2 = Saved from own harvest; 3 = Given by NGO; 4 = Given by NAADS/NGO working for NAADS/Private NAADS service provider; 5 = Given by government; 6 = Given by a friend/relative; 7 = Ginnery; 8 = Cooperative association, specify _____ ; 9 = Seed dealer; 10 = Trader not specializing in seed; 99 = Other (specify) _____

D.4 Please list inputs used for each parcel and plot for the COTTON SEASON OF 2007. (Please make sure the parcel and plot numbers correspond to the tables in D.2.)

Parcel name	Parcel ID#	Plot ID#	Use of herbicides and weeding					Use of chemical fertilizers									Use of pesticides					
			D.4.1 Used herbicide? Yes = 1 No = 0	D.4.2 Type of herbicide used (list on separate row if more than one)[1] If no, go to D.4.6	D.4.3 Amount used (kg/L)	D.4.4 Total value (USh)	D.4.5 Source of herbicide	D.4.6 Used chemical fertilizer? Yes = 1 No = 0 If no, go to D.4.11	D.4.7 Type of fert used (list on separate row if more than one)[2]	D.4.8 Amount used (kg)	D.4.9 Total value (USh)	D.4.10 Source of fert[3]	D.4.11 Used organic fertilizer? Yes = 1 No = 0 If no, go to D.4.15	D.4.12 Type of organic fertilizer[4]	D.4.13 Amount (kg)	D.4.14 Total value (USh)	D.4.15 Used pesticides / other chemicals? Yes = 1 No = 0 If no, go to D.4.20	D.4.16 Type of pesticide used (list on separate row if more than one)[5]	D.4.17 Amount used (kg)	D.4.18 Total value (USh)	D.4.19 Control which pest / disease?[6]	D.4.20 Total value of inputs used (USh)

[1]Type of herbicide: 1 = Roundup; 2 = Gramaxone; 3 = Mamba; 4 = Other (specify) _____. [2]Type of chemical fertilizers: 1 = NPK; 2 = Urea; 3 = CAN; 4 = SSP; 5 = Ammonium phosphate; 6 = DAP; 9 = Other (specify) _____. [3]Source of herbicide/chemical fertilizer/pesticide: 1 = Bought; 3 = Given by NGO not related to NAADS; 4 = Given by NAADS/NGO working for NAADS/Private NAADS service provider; 5 = Given by government; 6 = Given by a friend/relative; 99 = Other (specify) _____. [4]Type of organic fertilizer: 1 = Green manure; 2 = Animal manure; 3 = Compost; 99 = Other (specify) _____. [5]Type of pesticide: 1 = Biopesticides; 2 = Acheampong; 3 = Bt; 4 = Neem; 5 = Diapel; 6 = Biove; 9 = Extracts of Neem and Kyaya Grandifolia; 10 = Thionex; 11 = Nordox; 12 = Other, specify _____. [6]Pest/disease: 1 = Bollworm; 2 = Lygus bug; 3 = Aphids; 4 = Jassids; 5 = False coddling moth; 6 = Cotton stainers; 7 = Whitefly Bemisia; 8 = Spider mites; 9 = Bacterial blight; 10 = Wilt; 99 = Other (specify) _____.

D.5 Please list inputs used for each parcel and plot for the COTTON SEASON OF 2006. (Please make sure the parcel and plot numbers correspond to the tables in D.3.)

Parcel name	Parcel ID#	Plot ID#	Use of herbicides and weeding					Use of chemical fertilizers								Use of pesticides					
			D.5.1 Used herbicide? Yes = 1 No = 0	D.5.2 Type of herbicide used (list on separate row if more than one)[1] If no, go to D.5.6	D.5.3 Amount used (kg/L)	D.5.4 Total value (USh)	D.5.5 Source of herbicide	D.5.6 Used chemical fertilizer? Yes = 1 No = 0 If no, go to D.5.11	D.5.7 Type of fert used (list on separate row if more than one)[2]	D.5.8 Amount used (kg)	D.5.9 Total value (USh)	D.5.10 Source of fert[3]	D.5.11 Used organic fertilizer? Yes = 1 No = 0 If no, go to D.5.15	D.5.12 Type of organic fertilizer[4]	D.5.13 Amount (kg)	D.5.14 Total value (USh)	D.5.15 Used pesticides / other chemicals? Yes = 1 No = 0 If no, go to D.5.20	D.5.16 Type of pesticide used (list on separate row if more than one)[5]	D.5.17 Amount used (kg)	D.5.18 Total value (USh)	D.5.19 Control which pest / disease?[6]	D.5.20 Total value of inputs used (USh)

[1]Type of herbicide: 1 = Roundup; 2 = Gramaxone; 3 = Mamba; 4 = Other (specify) _____ . [2]Type of chemical fertilizers: 1 = NPK; 2 = Urea; 3 = CAN; 4 = SSP; 5 = Ammonium phosphate; 6 = DAP; 9 = Other (specify) _____ . [3]Source of herbicide/chemical fertilizer/pesticide: 1 = Bought; 3 = Given by NGO not related to NAADS; 4 = Given by NAADS/NGO working for NAADS; 5 = Given by government; 6 = Given by a friend/relative; 99 = Other (specify) _____ . [4]Type of organic fertilizer: 1 = Green manure; 2 = Animal manure; 3 = Compost; 99 = Other (specify) _____ . [5]Type of pesticide: 1 = Biopesticides; 2 = Acheampong; 3 = Bt; 4 = Neem; 5 = Diapel; 6 = Biovit; 9 = Extracts of Neem and Kyaya Grandifolia; 10 = Thionex; 11 = Nordox; 12 = Other, specify _____ . [6]Pest/disease: 1 = Bollworm; 2 = Lygus bug; 3 = Aphids; 4 = Jassids; 5 = False coddling moth; 6 = Cotton stainers; 7 = Whitefly Bemisia; 8 = Spider mites; 9 = Bacterial blight; 10 = Wilt; 99 = Other (specify) _____ .

D.6A   Cost of hired labor and family labor (cotton production)

Please report the family labor input in the COTTON SEASON OF 2007 for the cotton crop grown by this household. Report the number of days worked separately for adult females and males (16 years of age and above), and children (below the age of 16 years). (Please make sure the parcel and plot numbers correspond to the previous respective tables.)

Parcel name	Parcel ID#	Plot ID#	Activity[1]	Cost of hire, cotton season of 2007[2]	D.6a.1 Adult females			D.6a.2 Adult males			D.6a.3 Children (below 16 years)			D.6a.4 Labor donations		
					Number	Days worked	Average hours worked	Number	Days worked	Average hours worked	Number	Days worked	Average hours worked	Number	Days worked	Average hours worked

Please continue on the next page if needed

D.6A   Cost of hire and family labor input (cotton production) (continued)

Parcel name	Parcel ID#	Plot ID#	Activity[1]	Cost of hire, cotton season of 2007[2]	D.6a.1 Adult females			D.6a.2 Adult males			D.6a.3 Children (below 16 years)			D.6a.4 Labor donations		
					Number	Days worked	Average hours worked	Number	Days worked	Average hours worked	Number	Days worked	Average hours worked	Number	Days worked	Average hours worked

[1] 1 = Land clearing, 2 = First plough, 3 = Second plough, 4 = Planting, 5 = First fertilizer application, 6 = Second fertilizer application, 7 = Herbicide application, 8 = First pesticide application, 9 = Second pesticide application, 10 = Third pesticide application, 11 = First weeding, 12 = Second weeding, 13 = Third weeding, 14 = Harvesting, 15 = Post harvest handling, 16 = Transportation to markets, 99 = Other (specify) _____

[2] Includes cost of hiring the equipment: tractor, oxen ploughs, knapsack sprayer, etc. and the associated labor costs

D.6B   Cost of hire and family labor input (cotton production)

Please report the family labor input in the COTTON SEASON OF 2006 for the cotton crop grown by this household. Report the number of days worked separately for adult females and males (16 years of age and above), and children (below the age of 16 years). (Please make sure the parcel and plot numbers correspond to the previous respective tables.)

Parcel name	Parcel ID#	Plot ID#	Activity[1]	Cost of hire, cotton season of 2006[2]	D.6b.1 Adult females			D.6b.2 Adult males			D.6b.3 Children (below 16 years)			D.6b.4 Labor donations		
					Number	Days worked	Average hours worked	Number	Days worked	Average hours worked	Number	Days worked	Average hours worked	Number	Days worked	Average hours worked

Please continue on the next page if needed

D.6B Cost of hire and family labor input (cotton production) (continued)

Parcel name	Parcel ID#	Plot ID#	Activity[1]	Cost of hire, cotton season of 2006[2]	D.6b.1 Adult females			D.6b.2 Adult males			D.6b.3 Children (below 16 years)			D.6b.4 Labor donations		
					Number	Days worked	Average hours worked	Number	Days worked	Average hours worked	Number	Days worked	Average hours worked	Number	Days worked	Average hours worked

[1] 1 = Land clearing, 2 = First plough, 3 = Second plough, 4 = Planting, 5 = First fertilizer application, 6 = Second fertilizer application, 7 = Herbicide application, 8 = First pesticide application, 9 = Second pesticide application, 10 = Third pesticide application, 11 = First weeding, 12 = Second weeding, 13 = Third weeding, 14 = Harvesting, 15 = Post harvest handling, 16 = Transportation to markets, 99 = Other (specify) _____

[2] Includes cost of hiring the equipment: tractor, oxen ploughs, knapsack sprayer, etc. and the associated labor costs

D.7 Losses and marketable surplus

For all cotton cultivated, what proportion of output was marketed or lost during post harvest in the *Past 2 Cotton Seasons*? Also indicate any major catastrophic events since 2006, year of occurrence, and proportion of enterprise (pre-harvest) affected.

D.7.0 Prices over 3 year period (2005–2007; on the average):

lowest _____, highest _____, most frequent _____ (USh/kg)

Parcel name	Plot ID	D.7.1 Share marketed (%) If 0, go to D.7.7	D.7.2 Where sold* (list on separate row if more than 1 outlet was used)	D.7.3 Sales price (USh/ unit)	D.7.4 Unit	D.7.5 Who controls the sales' proceeds? 1 = Male 0 = Female Code	D.7.6 Relationship to household head	D.7.7 Storage losses (%)	D.7.8 Major source of storage losses**	D.7.9 Major event**	D.7.10 Year of occurrence	D.7.11 Proportion of enterprise affected (%)

Marketing — Losses — Major catastrophic events

*Codes for sales outlet: 1 = on farm, 2 = cooperative or farmer group, 3 = ginnery, 4 = bicycle traders, 5 = exporter, 6 = export market, 9 = other (specify) _____

**Codes for major events/source of storage loss: 0 = none, 1 = fire, 2 = flood, 3 = drought, 4 = major crop disease (specify) _____, major pest (specify) _____
99 = other (specify) _____

D.7.12 What are the major cotton marketing constraints? (Please list in order of importance)

1 _____

2 _____

3 _____

J   FARMER PERCEPTIONS

J.1   Seed

J.1.1 Why did you choose your primary seed source? _____ 1 = No other
choice, 2 = Good price, 3 = Good quality, 4 = Close, 5 = Trust the
source, 6 = Other (specify) _____

J.1.2 Apart from your primary source of seed, which other sources of seed do
you know of? _____

J.1.3 For how many years has the household grown cotton? _____

J.1.4 For how many years have you experienced bollworm? _____

J.1.5 Based on your experience over a number of years, please give us production levels for the following scenarios (show pictures):

Scenarios	Lowest production/acre		Highest production/acre		Most frequent production/acre	
	Original units	in kg	Original units	in kg	Original units	in kg
1) Absence of bollworm						
2) Presence of bollworm						
3) Presence of bollworm + pesticide						

J.1.6 If a cotton variety were resistant to bollworm so that you did not need to
apply much pesticide, would you be willing to pay, for this seed,

a. 50 percent more than what you paid last season? _____
(Yes = 1, No = 0)

b. If no, 25 percent more? _____ (Yes = 1, No = 0)

c. If no, 10 percent more? _____ (Yes = 1, No = 0)

J.1.7 If no for all the cases, what is the most you would pay for a cotton variety
which is resistant to cotton bollworm? (USh/kg) _____

J.1.8 For how many years have you had weed control problems? _____

J.1.9 Based on your experience over a number of years, please give us production levels for the following scenarios (show pictures):

Scenarios	Lowest production/acre		Highest production/acre		Most frequent production/acre	
	Original units	in kg	Original units	in kg	Original units	in kg
1) Presence of weeds (no manual weeding or herbicide)						
2) Presence of weeds + weeding (manual)						
3) Presence of weeds + herbicide						

J.1.10 If a cotton variety were resistant to Roundup so that you could control weeds with the cotton crop in the garden, would you be willing to pay for this seed:

    d. 50 percent more than what you paid last season? _____
      (Yes = 1, No = 0)

    e. If no, 25 percent more? _____ (Yes = 1, No = 0)

    f. If no, 10 percent more? _____ (Yes = 1, No = 0)

J.1.11 If no for all the cases, what is the most you would pay for a cotton variety which is Roundup Ready? (USh/kg) _____

J.2   Perceptions about Chemicals Use

  1. Do you apply yourself:

    pesticides (insecticides, fungicides, nematicides)? _____ (Yes = 1, No = 0)

    herbicides? _____ (Yes = 1, No = 0)

    If NO to both, go to Section E.1

  2. Which of the following have you experienced after applying chemicals? _____ 1. burning sensation on skin, 2. itchy or watery eyes, 3. very cold, 4. dizziness, 5. headache, 6. nausea or vomiting, 7. coughing, 8. breathing difficulties, 99. other (specify) _____

  3. Have any of the above been so severe that you have sought medical attention (including: conventional or traditional medicine) or self-medication? _____ (Yes = 1, No = 0)

  4. How many times, since you started cultivating cotton, have you sought medical attention (conventional, traditional, self-medication) after applying chemicals and suffering intoxication symptoms: _____

5. Did you receive training on how to use

   pesticides? _____ (Yes = 1, No = 0)

   herbicides? _____ (Yes = 1, No = 0)

   If yes, from whom? Specify _____

6. Do you put the chemicals in containers other than the originals?
   _____ (Yes = 1, No = 0)

7. Do you use empty pesticide containers for other uses? _____
   (Yes = 1, No = 0)

   If yes, specify _____

8. Where do you store your chemicals? _____

   Distance from house (meters) _____

9. Do you cover your mouth and nose when applying? _____
   (Yes = 1, No = 0)

10. Do you use gloves when applying chemicals? _____ (Yes = 1, No = 0)

11. Do you wear long sleeves and trousers or overalls when applying?
    _____ (Yes = 1, No = 0)

12. Do you wear gumboots when applying chemicals? _____
    (Yes = 1, No = 0)

13. Do you wear goggles? _____ (Yes = 1, No = 0)

14. Do you eat while applying chemicals? _____ (Yes = 1, No = 0)

15. Do you drink while applying chemicals? _____ (Yes = 1, No = 0)

16. Do you smoke while applying chemicals? _____ (Yes = 1, No = 0)

17. Do you wash after applying chemicals? _____ (Yes = 1, No = 0)

18. About pesticide use, do you think you use: _____ 1 = More, 2 = Less,
    3 = About the recommended amount of pesticide

19. About herbicide use, do you think you use: _____ 1 = More, 2 = Less,
    3 = About the recommended amount of herbicide

20. Do you understand the symbols that come on chemicals containers?
    _____ (Yes = 1, No = 0)

21. Do you read the instructions on the label? _____ (Yes = 1, No = 0)

E.1   Access to Cotton Advisory Services

E.1.1   In the PAST TWO COTTON SEASONS did you receive any advice/training from any service provider (agricultural extension services) for cotton production? _____ (Yes = 1, No = 0)

E.1.2   If NO to E.1.1, when was the last time you received advice/training on cotton production? _____ (month) _____ (year).

If YES to E.1.1, please report the frequency of visitation from various sources of extension services, mentioning the affiliation of the extension service provider(s)

Provider	Extension Code	E.1.3 Total number of visits in the PAST 2 COTTON SEASONS
NAADS service providers	1	
Government extension providers	2	
Farmer group member	3	
NGO not affiliated with government	4	
NGO but don't know affiliation	5	
Other farmers	6	
Project/program/volunteer providers	7	
Other (specify)		

E.2   Indicate availability of information and inputs in the PAST 2 COTTON SEASONS

NB: In this table production technology is used to designate the agricultural technology physical object/component (hardware) used in production, that is, variety, animal breed, ox-plough, post harvest equipment like maize shellers, etc., while production practice represents the knowledge/skills (software) required for optimal management and utilization of the agricultural technology physical object/component, that is, plant population and spacing, fertilizer application, disease control, etc.

	Available in community in the PAST 2 COTTON SEASONS? (Yes = 1, No = 0)
Information on	
E.2.1 Improved cotton production technologies	
E.2.2 Improved cotton production practices	
E.2.3 Market information (prices, markets, etc.) on cotton	
Physical availability of agricultural production inputs	
E.2.4 Improved seeds/planting material	
E.2.5 Inorganic fertilizers	
E.2.6 Pesticides/herbicides	
E.2.7 Farm equipment and tools	

Market Participation

E.2.8 Do you seek information about market prices before you plant? _____ (Yes = 1, No = 0)

E.2.9 Do you seek information about market preferences before you plant? _____ (Yes = 1, No = 0)

E.2.10 What are your key sources of information? _____ (1. Other farmers nearby (not in the association), 2. Other farmers in the association, 3. When I visit the market, 4. Radio or papers, 5. Non-governmental organizations operating in the area, 6. Government extension, 7. Ginnery, 99. Other, specify _____)

E.2.11 Are you able to obtain the information (e.g. channels, prices, preferences) you need for marketing your crop? _____ (Yes = 1, No = 0)

E.3 Access to Credit and Finance (e.g., inputs, agribusiness, assets, etc.) (consider all sources and both cash and in-kind)

E.3.1 Are you or is any member of this household aware of any credit available in the community in the past two years? _____ (Yes = 1, No = 0)

E.3.2 Did you or any member of this household apply/ask for any loan in the last two years? _____ (Yes = 1, No = 0). If Yes, go to E.3.4

E.3.3 If did not apply/ask, why not? _____
_____ Code _____

If did not apply, go to E.3.8

E.3.4 If applied/asked, for what purpose(s)? _____

E.3.5 If applied/asked, were any loans/credit received? _____ (Yes = 1, Partial = 2, No = 3)

E.3.6 If applied/asked but did not receive any loans/credit, why not?
_____ Code _____

E.3.7 If No or Partial to E.3.5, what did you do with your activity/plans? _____ (1 = abandoned the activity/plan, 2 = used own funds, 3 = got funds from other sources, 4 = still looking for other sources of funding, 5 = other (specify) _____)

E.3.8 In the past two years, did you or any member of your household receive any credit, grant, or subsidy? _____ (Yes = 1, No = 0). If No, go to section F

Amount and Terms of Any Credit or Subsidy Received Since 2006

For any credit or subsidy received SINCE 2006, indicate the amount from the government and other sources.

	Source 1	Source 2	Source 3	Source 4
E.3.9 Source of support				
Code*				
E.3.10 Nature of support (1 = credit, 2 = subsidy)				
E.3.11 Year received				
E.3.12 Purpose				
Code**				
E.3.13 Form of support				
Code***				
E.3.14 Total amount received in cash (USh)				
E.3.15 Amount/number of units received in-kind				
E.3.16 Units (e.g. kg, number, etc.)				
Unit code				
E.3.17 Total value of in-kind amount (USh at time of acquisition				
E.3.18 Was cash or in-kind amount received at the preferred time? (1 = Yes, 0 = No)				
E.3.19 Are there any repayment obligations? (1 = Yes, 2 = No)				
E.3.20 Payback modality (1 = cash, 2 = in-kind)				
E.3.21 Payback period (months)				
E.3.22 Interest rate on credit or payback amount (%)				
E.3.23 % repaid				
E.3.24 If there is no repayment obligation or repayment obligation is not 100%, why?				
Code				

*Source of credit/loan/grant/subsidy: 1 = Government program, 2 = NGO affiliated to government, 3 = Central/local government, 4 = NGO not affiliated to government, 5 = Friend/relative, 6 = Ginnery, 7 = MFI/SACCO, 8 = Politician, 9 = Other(specify), 10 = Bank

**Purpose: 1 = Fertilizer and/or other soil fertility management inputs, 2 = Seeds or planting materials, 4 = Other agricultural inputs, 5 = Other agricultural development activities

***Form of support: 1 = Improved seed, 2 = Livestock, 3 = Irrigation pump, 4 = Tractor, 5 = Housing, 6 = Oxen, 7 = Oxen plough, 8 = Cash, 99 = Other (specify)

F Membership in Institutions

F.1 Membership:

F.1.1 Does any person in your household belong to a cotton production interest group/association/cooperative?
___ (Yes = 1, No = 0). If No, go to Section G.

If Yes to F.1.1, what benefits have you realized from the most important economic interest groups SINCE 2004, if any? For each group, specify benefits and specify ease of household member in expressing his/her views in group decisionmaking.

F.1.2 Name of group to which you or any member of your household belongs	F.1.3 Year group was constituted	F.1.4 Major focus of group	Code	F.1.5 Role of member *	F.1.6 Major benefits realized **

*F.1.5 (role codes): 1 = lead farmer, 2 = chair or president, 3 = other executive member, 4 = member, 5 = secretary, 6 = salesman, 7 = other (specify)

**F.1.6 (benefits codes): 1 = access to extension advisory services, 2 = access to agricultural production inputs, 3 = exchange ideas, 4 = exchange labor, 5 = mobilize savings, 6 = get loans, 7 = collective marketing, 8 = improved standard of living, 9 = access to nonagricultural production inputs, 10 = uncertain about benefits, 11 = improved business, 12 = higher yields, 13 = acquired assets, 14 = not benefited, 15 = other (specify)

G   Household Income and Food Security

G.1   Household Expenditure

Please report the purchases done for consumption and non-consumption needs in the PAST 12 MONTHS

Cost item	Code	G.1.1 Monthly purchases (high expenditure season) (USh)	G.1.2 Number of months of high expenditure season	G.1.3 Monthly purchases (low expenditure season) (USh)	G.1.4 Number of months of low expenditure season
Food purchases	1	Annual expenditure			
Purchase of nonproductive durable goods	2				
Repair of houses and other durable assets plus maintenance	3				
Education	4				
Health	5				
Clothing and footwear	6				
Other (specify)					

G.2   Rank all sources of household incomes in terms of contribution to total household income in the past 12 months and indicate change in rank since 2006. NB: Question does not restrict itself to only cash income.

Income source	Income code	G.1.1 Rank in past 12 months[1,2]
Cotton	1	
All other crops	2	
Livestock	3	
Fishery	4	
Beekeeping	5	
Forestry	6	
Hunting	7	
Farm-labor wages	8	
Nonfarm**	9	

[1]Rank the sources, where 1 = first most important source, 2 = second most important, . . ., 9 = least important source.

[2]Codes for change: 1 = Increased a lot, 2 = increased a little, 3 = no change, 4 = decreased a little, 5 = decreased a lot

**Nonfarm income includes all sources of income other than agricultural production (i.e. remittances, brick-making, physical transfers, casual labor, salaried / wage labor, Bodaboda, trade, etc.)

END, THANK YOU FOR YOUR COOPERATION!!!

Experts Consulted in Uganda, 2007

Name	Position	Organization
Dr. Emmanuel Iyamuremye and Dr. Opolot Okasai	Employees	Crop Resources Department, Ministry of Agriculture, Animal Industries and Fisheries
Dr. Arthur Makara	Biosafety officer	Uganda National Council of Science and Technology
Ms. Jolly K. Sabune	Executive director	Cotton Development Organisation
Mr. Hans Muzoora	Principal market information and monitoring officer	
Mr. Damalie Lubwana	Agronomist	
Ms. Jane Nalunga	International monitoring officer	National Organic Agricultural Movement of Uganda
Mr. Patrick Oryang	Ginner, certified-organic cotton lint	Lango Cooperative Union (Ngetta Ginnery)
Mr. Dennis Kaijabahoire and Mr. Adam Bwambale	Members	Nyakatonzi Cooperative Union
Dr. J. Mwesigwa Magyembe	Special assistant to the director general	National Agricultural Research Organisation
Dr. Annunciata Hakuza	Senior agricultural economist	
Dr. Sunday Godfrey	Statistician	
Dr. Thomas E. E. Areke	Officer in charge of the Bt cotton confined field trials	
Mrs. Florence Kata	Executive director	Uganda Export Promotions Board
Mr. Amos Tindyobwa	Director of market research	
Mr. Othieno Odoi	Senior trade promotion officer	
Dr. Tila Zeweldu Abebe	Biotechnology advisor	Agricultural Productivity Enhancement Project
Mr. Herbert Kirunda	Commercialization specialist–cotton	
Dr. Serunjogi Lastus Katende	Chairman of the board; cotton breeder, former director of Serere Agricultural and Animal Research Institute	Cotton Development Organisation

Source: Authors.

Ginneries Report on Seed Cotton and Cotton Lint, October 7, 2007

Ginnery company	Districts with ginneries	Seed cotton (kilograms)	Lint (kilograms)	Share of total lint production (percent)
Dunavant	Iceme, Odokomit, Kabulubulu, N'gola, Kitgum, Ngetta	11,381,222	4,080,908	16.57
Pramukh Agro/Bushenyi Cotton	Busembatia, Ladoto, Bulumba/ Bushenyi	6,669,543	2,347,577	9.71
Copcot	Parombo, R/Camp, Masindi	5,710,045	2,041,605	8.31
North Bukedi Cotton Company	Iki Iki, Kabole, Bugema	5,365,561	1,851,472	7.81
Western Uganda Cotton Company	Kasese, Masindi, Pakwach	5,002,014	1,771,984	7.28
Olam	Kibuku, Hoima, Parombo, Pakwach	4,572,820	1,669,942	6.66
CN Cotton/Kumi Cotton	Kachumbala/Mukhongoro	4,561,504	1,614,373	6.64
LOFP/Bollweevil	Spinning Mill	3,362,266	1,162,401	4.90
South Base Agro	Busolwe, Jaber	2,705,464	958,462	3.94
Nyakatonzi Co-op Union	Kasese	2,694,596	912,800	3.92
Rwenzori Ginners	Rwenzori	2,599,237	890,546	3.78
Bon Holdings	Bulumba, Nakivumbi, Luzinga	2,542,892	897,099	3.70
Novo Enterprises	Nyakesi	1,840,456	645,510	2.68
Uganda Cotton Klub	Bulangira	1,766,087	577,094	2.57
Bestlines	Dabani	1,747,102	609,155	2.54
Twin Brothers	Aduku	1,657,379	613,354	2.41
Mutuma Commercial Agencies	Kiyunga	667,145	239,685	0.97
Masaba Cotton	Lukhonge	646,898	224,330	0.94
Country Farm	Soroti	645,413	243,344	0.94
Balawoli Ginners	Balawoli	606,153	212,353	0.88
Lango Cooperative Union	Ngetta	565,866	202,461	0.82
Rafiki Cotton Industries	Aboke	426,160	147,885	0.62
East Acholi Coop Union	Kitgum	381,313	137,061	0.56
Intraco	Busembatia	278,419	103,323	0.41
Nyamambuka Farmers	Kasese Union	101,794	33,871	0.15
Kakyu Hardware	Rwenzori	97,120	31,685	0.14
FICA	Masindi	87,000	33,277	0.13
Total		68,681,469	24,253,557	100

Sources: S. Godfrey, Ugandan Ministry of Agriculture, National Agricultural Research Organisation (personal communication, 2007); CDO (2006).

History of Organic Cotton in Uganda

In 1994 the organization Export Promotion of Organic Products from Africa began the Lango Organic Project in the Lira and Apac districts. The project was promoted by Bo Weevil BV, a Dutch trader in organic textiles, as a business-oriented enterprise rather than an environmentally oriented project (Tulip and Ton 2002). The region is rich in black ants (*ngini ngini*), natural predators for cotton pests (Moseley and Gray 2008).

In recent years, organic production has expanded to other crops besides cotton. According to Taylor (2006) there were almost 38,000 certified farmers in Uganda in 2004, almost one-third of them producing cotton. From having just a few private companies promoting the production of organic cotton, the organic sector in Uganda has grown to include multiple actors (Table A4.1) and crops. The National Organic Agricultural Movement of Uganda (NOGAMU) was established in 2001 and by mid-2005 had attracted more than 300 individual members and 80 corporate members. Currently NOGAMU is linked to 25,000 stakeholders in the organic sector.[1]

As more farmers and organizations have become involved in organic farming, standards to govern such production have also emerged. NOGAMU promoted the Uganda Organic Standards and established UgoCert in 2004 to certify that products have been produced according to specified standards. Certification carried out by UgoCert is based on third-party inspections. To be able to export, international companies must certify that products exported comply with export country standards. Ecocert and the Institute of Market Ecology are the international companies that verify this compliance for organic cotton exports to the European market.

---

[1] Organic production in Uganda has had some setbacks, however. Tulip and Ton (2002) report that the Lango Project attracted a lot of attention from private companies and donors, but it seems to have expanded beyond its financial and marketing capabilities. The same authors point out that some of the original organic producers have abandoned the project, discouraged because only a minor percentage of the certified-organic production is bought from farmers at organic premiums.

TABLE A4.1   Institutions involved in the production of organic cotton

Local	National	Regional	International
	Organic-cotton-related organizations		
Lango Organic Project (1994)	National Organic Agricultural Movement of Uganda (2001)	Africa Organic Center (2004)	International Federation of Organic Agricultural Movements (2004)
Northern Uganda Eco Organic (2004)	UgoCert (2004) Uganda Organic Standards	Export Promotion of Organic Products from Africa (1994)	Bo Weevil BV
Four ginneries (two local and two owned by Dunavant)			
	Other organizations		
Other ginneries and buyers	Uganda Export Promotion Board		

Source: Author's elaboration based on secondary information available (P. Zambrano).
Note: Years in parentheses are when the organizations were established.

Also involved in setting organic standards is the International Federation of Organic Agricultural Movements (IFOAM), which established the IFOAM Africa Office in 2004. IFOAM has been collaborating in the establishment of organic agricultural standards for East Africa, an initiative of the United Nations Environmental Program–United Nations Conference on Trade and Development. It has also developed the Participatory Guarantee Systems.

Dunavant, the largest cotton company in Uganda, also took an interest in organic cotton and, along with the cotton companies the Lango Cooperative Union and Outspan Enterprises, established Northern Uganda Eco Organic in 2004 to handle organic cotton. Of the six ginneries that Dunavant operates, Odokomit and Ngetta in Lira district were certified in July 2006. These are the only certified-organic ginneries in Uganda.[2]

Northern Uganda Eco Organic establishes agreements with organic farmer groups. Under these agreements, the ginners supply farmers with seed and farm tools such as hoes, slashers, and pangas (machetes). They also offer extension services to farmers and commit to buying cotton from these farmers. In exchange, farmers agree to follow specific practices recommended by the organic ginneries; if the conditions are not met, farmers receive a reduced price. To what extent these agreements are followed is yet to be evaluated.

---

2 For more on these ginneries, see www.dunavant.com/Offices/Geneva/OrganicCotton/tabid/152/Default.aspx.

Net-Map Toolbox

Net-Map was developed by Eva Schiffer (Schiffer 2007; Schiffer, Narrod, and Grebner 2008) to help "people understand, visualize, discuss, and improve situations in which many different actors influence outcomes. . . . By creating Influence Network Maps, individuals and groups can clarify their own view of a situation, foster discussion, and develop a strategic approach to their networking" (Net-Map Toolbox: Influence Mapping of Social Networks 2013).

In the case of Uganda, the study used the Net-Map tool to clarify the regulatory process and to identify strategic actors and agencies and bottlenecks in the approval process of both the Biosafety Bill and the implementation of confined field trials.

No methodology for assessing the regulatory system and determining how long it will take to approve a technology is foolproof. Nevertheless, Net-Map helps this evaluation process by making it possible to visually map all agents involved in the institutional framework, document regulatory activity, assess the relative influence of each agent, and identify crucial bottlenecks. This study used Net-Map as a tool for this purpose and also to identify all stakeholders involved in the cotton production sector and the main links among them.

The successful implementation of the Net-Map approach requires the clear definition of research questions to frame the exercise. In the case of GM cotton in Uganda, the researchers were interested in answering two questions: Which institutions and agents were in charge of the approval of confined cotton field trials, and what were the bottlenecks in this process? Who were the main actors in the approval process of the Biosafety Bill in Uganda? It is important to note that the field trials have been approved since this exercise took place, but the approval process gives a clear picture of where the process had encountered bottlenecks. This is not the case regarding the approval of the Biosafety Bill, which, despite being drafted before the study began, has yet to be approved and continues to stagnate.

Network mapping should be implemented on site with the participation of all relevant actors. Because of budget and time constraints, this study was unable to involve all relevant actors and had to limit the sources of information for this exercise to secondary information, expert consultation, and small-group discussions, complemented by data collected by Ugandan study partners. Using this available information and Net-Map techniques, the study limited its analysis to documenting and mapping the regulatory process and identifying all agents in the delivery of the genetically modified technology to cotton farmers. Ideally, this mapping of the regulatory process would need to be validated by and discussed among the different mapped stakeholders, but such stakeholder involvement requires additional resources.

Stochastic Dominance

The first-degree stochastic dominance rules use the assumption that the decisionmaker's expected utility function expresses only positive marginal utility: that is, the function reflects that for every unit of a good consumed, the consumer receives some additional utility. For two alternatives A and B (Bt cotton and conventional cotton, for example), A is first-degree stochastic dominant over B if for all x with at least one strict inequality the following holds:

$$F_A(x) \geq F_B(x),$$

where $F_A(x)$ is the cumulative distribution function of alternative A, and $F_B(x)$ is the cumulative distribution function of alternative B. In this study's case, x is the change in net benefits for each scenario simulated using partial-budget (Chapter 5) or economic surplus (Chapter 6) methods.

This assumption implies that for all values of x in the sample, one cumulative distribution function is always equal to or greater than that of the other, and therefore the functions do not cross. If this is not true, then additional assumptions have to be made to allow the possibility of ranking these functions based on which is greater than the other for more values x. In the case of second-degree stochastic dominance, the additional assumption is that for the decisionmaker's expected utility of risk aversion, the decisionmaker's expected utility function is positive but has a decreasing slope for all values of x. Thus, for two alternatives A and B, A is second-degree stochastic dominant over B if for all x^* with at least one strict inequality, one has

$$\int_{-\infty}^{x^*} F_A(x)dx \leq F_B(x)dx$$

The assumption of the decisionmaker being risk averse is more powerful yet restrictive, as it is necessary to express the degree of risk averseness. In practice this value may be hard to estimate, or the estimation may not be feasible. However, farmers in developing countries are assumed to be risk averse. Thus, second-degree stochastic dominance rules have to be fine tuned to adjust and better model the risk-averse nature of farmers. Note that with these rules, one is in essence considering the cumulative distribution under the density functions, and thus simultaneously assessing mean values of the parameters and the probability of risk.

Sensitivity Analysis of Marginal Benefits of the Organic Production Plus Premium Price Simulation

FIGURE A7.1   Sensitivity analysis of marginal benefits

Organic + premium price
Regression coefficient

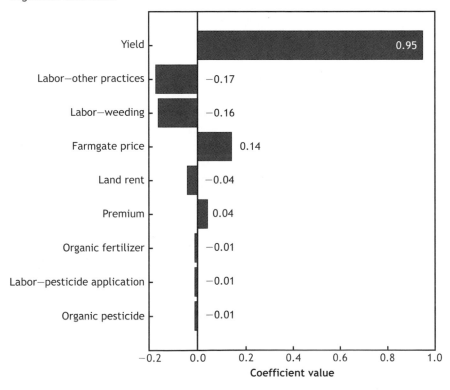

Coefficient value

Source: Authors.

Summary of the Net Present Value of Net Benefits from Economic Surplus Estimations

Statistic	Scenario I: Public-sector release	Scenario II: Private-sector release with reduced technology fee	Scenario III: Private-sector release with reduced technology fee	Scenario IV: Both insect resistance and herbicide tolerance with reduced technology fee	Scenario V: Both insect resistance and herbicide tolerance with reduced technology fee	Scenario VI: Dual trait with reduced technology fee and increased total cotton area planted and adoption rate
Mean	6,804,135	1,533,175	4,121,746	1,821,044	7,473,349	34,104,934
Coefficient of variation	82.5	217.5	100.1	193.1	99.2	157.6
Minimum	−13,542,218	−15,319,684	−13,751,645	−10,861,621	−17,197,714	−80,723,708.64
Maximum	58,461,170	40,905,302	48,722,367	44,400,004	79,653,243	726,878,700
Standard deviation	5,612,472	3,382,096	4,125,180	3,516,164	7,416,912	53,732,520
Skewness	1.41	2.15	1.66	2.14	1.66	3.28
Kurtosis	7.57	12.14	8.81	12.17	9.13	20.70
Median	5,789,850	654,567	3,285,597	986,448	6,087,736	16,814,823
Mode	3,637,683	−379,544	2,761,927	−771,188	3,998,883	−2,778,932

Source: Authors.
Note: All values are US dollars.

Summary of Total Net Benefits Statistics from Economic Surplus Estimations

Statistic	Scenario I: Public-sector release	Scenario II: Private-sector release with technology fee	Scenario III: Private-sector release with reduced technology fee	Scenario IV: Both insect resistance and herbicide tolerance with technology fee	Scenario V: Both insect resistance and herbicide tolerance with reduced technology fee	Scenario VI: Dual trait with reduced technology fee and increased total cotton area planted and adoption rate
Mean	64,997,970	15,989,645	39,961,923	18,626,829	71,299,556	319,700,453
Minimum	−108,881,848	−162,743,338.85	−109,137,723	−106,104,400	−152,970,700	−743,377,766
Maximum	547,326,578	326,514,667	366,759,615	340,239,961	675,685,886	6,780,369,129
Standard deviation	51,700,235	30,448,199	37,651,899	31,766,696	67,739,024	501,365,898
Skewness	1.36	2.08	1.57	2.06	1.57	3.28
Kurtosis	7.27	11.40	7.99	11.29	8.50	20.79
Median	56,233,053	8,339,296	32,563,922	11,509,217	59,265,859	158,351,296
Mode	34,783,137	1,619,989	22,825,748	13,462,376	32,040,353	−12,752,173

Source: Authors.
Note: All values are US dollars.

Summary of Internal Rate of Return Statistics from Economic Surplus Estimations

Statistic	Scenario I: Public-sector release	Scenario II: Private-sector release with technology fee	Scenario III: Private-sector release with reduced technology fee	Scenario IV: Both insect resistance and herbicide tolerance with technology fee	Scenario V: Both insect resistance and herbicide tolerance with reduced technology fee	Scenario VI: Dual trait with reduced technology fee and increased total cotton area planted and adoption rate
Mean (%)	81	54	68	56	83	129
Minimum (%)	−3	−10	−6	0	1	1
Maximum (%)	172	163	161	168	202	309
Standard deviation (%)	23	27	23	27	26	47
Skewness	−0.1400	0.1600	0.0005	0.1200	−0.0900	0.1200
Kurtosis	3.33	2.60	3.00	2.61	3.24	2.88
Median (%)	82	54	69	56	84	128
Mode (%)	89	51	69	64	78	139

Source: Authors.

References

ACE (Audit Control & Expertise (U) Ltd). 2006. "End of Cotton Season Report 2005/06." Mimeo, submitted to Uganda Ginners and Cotton Exporters Association, Kampala, Uganda.

Alexander, C., and R. E. Goodhue. 2002. "The Pricing of Innovations: An Application to Specialized Corn Traits." *Agribusiness* 18 (3): 333–348.

Alston, J. M., G. W. Norton, and P. G. Pardey. 1995. *Science under Scarcity: Principles and Practice for Agricultural Research Evaluation and Priority Setting.* Ithaca and London: Cornell University Press.

Anderson, K., E. Valenzuela, and L. A. Jackson. 2008. "Recent and Prospective Adoption of Genetically Modified Cotton: A Global Computable General Equilibrium Analysis of Economic Impacts." *Economic Development and Cultural Change* 56 (2): 265–296.

APSEC (Agricultural Policy Secretariat). 1998. *Report on Economics of Crop and Livestock Production, Processing and Marketing.* Kampala, Uganda.

———. 2001. *An Assessment of the Impact of Cotton Sub-Sector Development Project (CSDP).* Kampala, Uganda.

Areal, F. J., L. Riesgo, and E. Rodriguez-Cerezo. 2012. "Economic and Agronomic Impact of Commercialized GM Crops: A Meta-Analysis." *Journal of Agricultural Science* 151 (1): 1–27.

Ayer, H. W., and G. E. Schuh. 1972. "Social Rates of Return and Other Aspects of Agricultural Research: The Case of Cotton Research in Sao Paulo, Brazil." *Journal of Agricultural Economics* 54 (4 pt.1): 557–569.

Baffes, J. 2009. "The 'Full Potential' of Uganda's Cotton Industry." *Development Policy Review* 27 (1): 67–85.

Bennett, R., Y. Ismael, and S. Morse. 2005. "Explaining Contradictory Evidence Regarding Impacts of Genetically Modified Crops in Developing Countries. Varietal Performance of Transgenic Cotton in India." *Journal of Agricultural Science* 143 (1): 35–41.

Bennett, R., Y. Ismael, U. Kambhampati, and S. Morse. 2004. "Economic Impact of Genetically Modified Cotton in India." *AgBioForum* 7 (3): 96–100.

Bennett, R., U. Kambhampati, S. Morse, and Y. Ismael. 2006. "Farm-Level Economic Performance of Genetically Modified Cotton in Maharashtra, India." *Review of Agricultural Economics* 28 (1): 59–71.

Birol, E., E. R. Villalba, and M. Smale. 2008. "Farmer Preferences for Milpa Diversity and Genetically Modified Maize in Mexico: A Latent Class Approach." *Environment and Development Economics* 14 (4): 521–540.

Bouët, A., and G. Gruère. 2011. "Refining Estimates of the Opportunity Cost of Non-Adoption of Bt Cotton: The Case of Seven Countries in Sub-Saharan Africa." *Applied Economic Perspectives and Policy* 33 (2): 260–279.

Carr, S. 1993. *Improving Cash Crops in Africa: Factors Influencing the Productivity of Cotton, Coffee, and Tea Grown by Smallholders*. World Bank Technical Paper 216. Washington, DC: World Bank.

CDO (Cotton Development Organisation). 2004. "Cotton Development Organisation Annual Report 2003–2004." Mimeo, Kampala, Uganda.

———. 2006. "The Cotton Sector in Uganda: Progress Made in the Sector and Recommendations for Achieving Further Progress." Mimeo, Kampala, Uganda.

———. 2008. "Cotton Development Organisation Annual Report 2007/08." Mimeo, Kampala, Uganda.

CIMMYT (International Center for Maize and Wheat Improvement). 1988. *From Agronomic Data to Farmer Recommendations: An Economics Training Manual*. Mexico, D.F.

CPB (Cartagena Protocol on Biosafety). 2000. "The Cartagena Protocol on Biosafety." Accessed May 20, 2013. http://bch.cbd.int/protocol/.

Davis, G. C., and M. C. Espinoza. 1998. "A Unified Approach to Sensitivity Analysis in Equilibrium Displacement Models." *American Journal of Agricultural Economics* 80 (4): 868–879.

Delgado, C., and N. Minot. 2000. *Agriculture in Tanzania since 1986: Follower or Leader of Growth?* Washington, DC: World Bank and International Food Policy Research Institute.

Demont, M., M. Cerovska, W. Deams, K. Dillen, J. Fogarasi, E. Mathijs, F. Muska, et al. 2008. "Ex ante Impact Assessment under Imperfect Information: Biotechnology in New Member States of the EU." *Journal of Agricultural Economics* 59 (3): 463–486.

Dercon, S. 1993. "Peasant Supply Response and Macroeconomic Policies: Cotton in Tanzania." *Journal of African Economics* 2 (2): 157–193.

Dyer, G. A., J. A. Serratos-Hernández, H. R. Perales, P. Gepts, A. Piñeyro-Nelson, A. Chávez, N. Salinas-Arreourtua, et al. 2009. "Dispersal of Transgenes through Maize Seeds Systems in Mexico." *PLoS ONE* 4 (5): e5734. doi:10.1371/journal.pone.0005734.

Edmeades, S., and M. Smale. 2006. "A Trait-Based Model of the Potential Demand for a Genetically Engineered Food Crop in a Developing Economy." *Agricultural Economics* 35 (3): 351–361.

Elbehri, A., and S. MacDonald. 2004. "Estimating the Impact of Transgenic Bt-Cotton on West and Central Africa: A General Equilibrium Approach." *World Development* 32 (12): 2049–2064.

Falck-Zepeda, J. 2006. "Coexistence, Genetically Modified Biotechnologies, and Biosafety: Implications for Developing Countries." *American Journal of Agricultural Economics* 88 (5): 1200–1208.

Falck-Zepeda, J., and J. I. Cohen. 2003. "Accessing Agricultural Biotechnology in Emerging Economies." In *Proceedings of a Workshop on Impact Assessment and Agricultural Biotechnology—Research Methodologies for Developing, Emerging and Transition Economies,* 47–71. Paris: Organisation for Economic Co-operation and Development.

Falck-Zepeda, J., and P. Zambrano. 2011. "Socioeconomic Considerations in Biosafety and Biotechnology Decision Making: The Cartagena Protocol and National Biosafety Frameworks." *Review of Policy Research* 28 (2): 171–195.

Falck-Zepeda, J., J. D. Horna, and M. Smale. 2008. "Betting on Cotton: Potential Payoffs and Economic Risks of Adopting Transgenic Cotton in West Africa." *African Journal of Agricultural and Resource Economics* 2 (2): 188–207.

Falck-Zepeda, J. B., G. Traxler, and R. G. Nelson. 2000. "Surplus Distribution from the Introduction of a Biotechnology Innovation." *American Journal of Agricultural Economics* 82 (2): 360–369.

Falck-Zepeda, J., J. Wesseler, and S. Smyth. 2010. "The Current Status of the Debate on Socioeconomic Assessments and Biosafety: Highlighting Different Positions and Policies in Canada and the US, the EU and Developing Countries." Paper presented at the World Environmental and Resource Economics Congress, Montreal, July 2.

FAO (Food and Agriculture Organization of the United Nations). 2010. FAOSTAT. Production and Trade Data. Accessed October 2010. http://faostat.fao.org/.

Feder, G., and D. Umali. 1993. "The Adoption of Agricultural Innovation: A Review." *Technological Forecasting and Cultural Change* 43 (3–4): 215–239.

Feder, G., R. E. Just, and D. Zilberman. 1985. "Adoption of Agricultural Innovations in Developing Countries: A Survey." *Economic Development and Cultural Change* 33 (2): 255–298.

Finger, R., N. E. Benni, T. Kaphengst, C. Evans, S. Herbert, B. Lehmann, S. Morse, and N. Stupak. 2011. "A Meta-Analysis on Farm-Level Costs and Benefits of GM Crops." *Sustainability* 3 (5): 743–762.

Fisher, M. G., W. A. Masters, and M. Sidibe. 2001. "Technical Change in Senegal's Irrigated Sector: Impact Assessment under Uncertainty." *Agricultural Economics* 24 (2): 179–197.

Fok, M., W. Liang, and Y. Wu. 2005. "Diffusion du coton génétiquement modifié en Chine: Leçons sur les facteurs et limites d'un succès." *Economie rurale* 285 (January–February): 5–32.

Frisvold, G. B., J. M. Reeves, and R. Tronstad. 2006. "Bt Cotton Adoption in the United States and China: International Trade and Welfare Effects." *AgBioForum* 9 (2): 69–78.

Gergely, N., and C. Poulton. 2009. "Historical Background and Recent Evolution of African Cotton Sectors." In *Organization and Performance of Cotton Sectors in Africa, Learning from Reform Experience*, edited by D. Tschirley, C. Poulton, and P. Labaste, 31–44. Washington, DC: World Bank.

Glover, D. 2010. "Is Bt Cotton a Pro-Poor Technology? A Review and Critique of the Empirical Record." *Journal of Agrarian Change* 10 (4): 482–509.

Gordon, A., and A. Goodland. 2000. "Production Credit for African Smallholders: Conditions for Private Provision." *Savings and Development* 24 (1): 55–84.

Gouse, M., J. Kirsten, B. Shankar, and C. Thirtle. 2005. "Bt Cotton in KwaZulu Natal: Technological Triumph but Institutional Failure." *AgBiotechNet* 7 (134): 1–7.

Gruère, G., and M. Cartel. 2007. "Introduction of Bt Cotton into Cotton Market Channels." In *Assessing the Potential Economic Impact of Bt Cotton in West Africa: Preliminary Finding and Elements of a Proposed Methodology,* edited by M. Smale, G. Gruère, J. Falck-Zepeda, A. Bouët, J. D. Horna, M. Cartel, and P. Zambrano, 45–84. Project report for the World Bank, Washington, DC.

Gruère, G., and D. Sengupta. 2009. *Biosafety and Perceived Commercial Risks: The Role of GM Free Private Standards.* Discussion Paper 00847. Washington, DC: International Food Policy Institute.

Gruère, G., A. Bouët, and S. Mevel. 2007. *Genetically Modified Food and International Trade: The Case of Bangladesh, India, Indonesia and the Philippines.* Discussion Paper 740. Washington, DC: International Food Policy Institute.

Gruère, G., P. Mehta-Bhatt, and D. Sengupta. 2008. *Bt Cotton and Farmer Suicides: Reviewing the Evidence.* Discussion Paper 00808. Washington, DC: International Food Policy Research Institute.

Hardaker, J. B., R. B. M. Huirne, J. R. Anderson, and G. Lien. 2004. *Coping with Risk in Agriculture.* Wallingford, UK: CABI Publishing.

Hareau, G. G., B. F. Mills, and G. W. Norton. 2006. "The Potential Benefits of Herbicide Resistant Transgenic Rice in Uruguay: Lessons for Small Developing Countries." *Food Policy* 31 (2): 162–179.

Horna, J. D., M. Smale, R. Al-Hassan, J. Falck-Zepeda, and S. E. Timpo. 2008. *Insecticide Use on Vegetables in Ghana: Would GM Seed Benefit Farmers?* Discussion Paper 00785. Washington, DC: International Food Policy Research Institute.

Huang, J., R. Hu, S. D. Rozelle, F. Qiao, and C. E. Pray. 2001. *Smallholders, Transgenic Varieties, and Production Efficiency: The Case of Cotton Farmers in China.* Department of Agricultural and Resource Economics Working Paper 01-015. Davis, CA, US: University of California, Davis. SSRN: http://ssrn.com/abstract=321761 or doi:10.2139/ssrn.321761.

Huang, J., R. Hu, C. Fan, C. Pray, and S. Rozelle. 2002a. "Bt Cotton Benefits, Costs, and Impacts in China." *AgBioForum* 5 (4): 153–166.

Huang, J., R. Hu, S. Rozelle, F. Qiao, and C. E. Pray. 2002b. "Transgenic Varieties and Productivity of Smallholder Cotton Farmers in China." *Australian Journal of Agricultural and Resource Economics* 46 (3): 367–387.

Huang, J., R. Hu, Q. Wang, J. Keeley, and J. Falck-Zepeda. 2002c. "Agricultural Biotechnology Development, Policy and Impact." *Economic and Political Weekly* 37 (27): 2756–2761.

Huang, J., R. Hu, C. Pray, F. Qiao, and S. Rozelle. 2003. "Biotechnology as an Alternative to Chemical Pesticides: A Case Study of Bt Cotton in China." *Agricultural Economics* 29 (1): 55–67.

Huang, J., R. Hu, H. van Meijl, and F. van Tongeren. 2004. "Biotechnology Boost to Crop Productivity in China: Trade and Welfare Implications." *Journal of Development Economics* 75 (1): 27–54.

Huygen, I., M. Veeman, and M. Lehrol. 2003. "Cost Implications of Alternative GM Tolerance Levels: Non-Genetically Modified Wheat in Canada." *AgBioForum* 6 (4): 169–177.

ICAC (International Cotton Advisory Committee). 2008. World Cotton Trade 2008. Washington, DC.

———. 2010. World Cotton Trade 2010. Washington, DC.

IFPRI (International Food Policy Research Institute) and HarvestChoice. 2012. "MAPPR map tool." Accessed May 14, 2013. harvestchoice.org/mappr.

James, C. 2007. *Global Status of Commercialized Biotech/GM Crops 2007*. Brief 37. Ithaca, NY, US: International Service for the Acquisition of Agri-biotech Applications.

———. 2008. *Global Status of Commercialized Biotech/GM Crops 2008*. Brief 39. Ithaca, NY, US: International Service for the Acquisition of Agri-biotech Applications.

———. 2009. *Global Status of Commercialized Biotech/GM Crops 2009*. Brief 41. Ithaca, NY, US: International Service for the Acquisition of Agri-biotech Applications.

James, C. 2010. *Global Status of Commercialized Biotech/GM Crops 2010.* Brief 42. Ithaca, NY, US: International Service for the Acquisition of Agri-biotech Applications.

———. 2011. *Global Status of Commercialized Biotech/GM Crops 2011.* Brief 43. Ithaca, NY, US: International Service for the Acquisition of Agri-biotech Applications.

———. 2012. *Global Status of Commercialized Biotech/GM Crops 2012.* Brief 44. Ithaca, NY, US: International Service for the Acquisition of Agri-biotech Applications.

Kabwe, S., and D. Tschirley. 2007. "Farm Yields and Returns to Farmers from Seed Cotton: Does Zambia Measure Up?" Policy Synthesis: Food Security Research Project—Zambia, No. 26. Accessed May 21, 2013. http://fsg.afre.msu.edu/zambia/ps26.pdf.

Kathage, J., and M. Qaim. 2012. "Economic Impacts and Impact Dynamics of Bt (*Bacillus thuringiensis*) Cotton in India." *Proceedings of the National Academy of Sciences USA* 109 (29): 11652–11656.

Kelly, V. 2006. *Factors Affecting the Demand for Fertilizer in Sub-Saharan Africa.* Agriculture and Rural Development Discussion Paper 23. Washington, DC: World Bank.

Kikulwe, E. 2010. "On the Introduction of Genetically Modified Bananas in Uganda: Social Benefits, Costs, and Consumer Preferences." Dissertation, Wageningen University, Wageningen, the Netherlands.

Kikulwe, E., E. Birol, J. Wesseler, and J. Falck-Zepeda. 2009. *A Latent Class Approach to Investigating Consumer Demand for Genetically Modified Food in a Developing Country: The Case of GM Bananas in Uganda.* Discussion Paper 00938. Washington, DC: International Food Policy Research Institute.

Kolady, D. E., and W. Lesser. 2006. "Who Adopts What Kind of Technologies? The Case of Bt Eggplant in India." *AgBioForum* 9 (2): 94–103.

———. 2008a. "Is Genetically Engineered Technology a Good Alternative to Pesticide Use? The Case of GE Eggplant in India." *International Journal of Biotechnology* 10 (2/3): 132–147.

———. 2008b. "Potential Welfare Benefits from the Public-Private Partnerships: A Case of Genetically Engineered Eggplant in India." *Journal of Food Agriculture and Environment* 6 (3/4): 333–340.

———. 2009. "Can Owners Afford Humanitarian Donations in Agbiotech? The Case of Genetically Engineered Eggplant in India." *Electronic Journal of Biotechnology* 11 (2): 1–8.

Krishna, V. V., and M. Qaim. 2007. "Estimating Adoption of Bt Eggplant in India: Who Benefits from Public-Private Partnerships?" *Food Policy* 32 (2): 523–543.

Lichtenberg, E., and D. Zilberman. 1986. "The Econometrics of Damage Control: Why Specifications Matter." *American Journal of Agricultural Economics* 68 (2): 261–273.

Lin, W. 2002. "Estimating the Costs of Segregation for Non-Biotech Maize and Soybeans." In *Market Development for Genetically Modified Foods*, edited by V. Santaniello, R. E. Evenson, and D. Zilberman, 261–265. Wallingford, UK: CABI Publishing.

Lybbert, T. J., and A. Bell. 2010. "Stochastic Benefit Streams, Learning, and Technology Diffusion: Why Drought Tolerance Is Not the New Bt." *AgBioForum* 13 (1): 13–24.

Meinzen-Dick, R., A. Quinsumbing, J. Behrman, P. Biermayr-Jenzano, V. Wilde, M. Noordeloos, C. Ragasa, and N. Beintema. 2010. *Engendering Agricultural Research*. Discussion Paper 00973. Washington, DC: International Food Policy Research Institute.

Minot, N., and L. Daniels. 2005. "Impact of Global Cotton Markets on Rural Poverty in Benin." *Agricultural Economics* 33 (3): 453–466.

Morse, S., R. Bennet, and Y. Ismael. 2005a. "Genetically Modified Insect Resistance in Cotton: Some Farm Level Economic Impacts in India." *Crop Protection* 24 (5): 433–440.

———. 2005b. "Comparing the Performance of Official and Unofficial Genetically Modified Cotton in India." *AgBioForum* 8 (1): 1–6.

———. 2007. "Isolating the 'Farmer' Effect as a Component of the Advantage of Growing Genetically Modified Varieties in Developing Countries: A Bt Cotton Case Study from Jalgaon, India." *Journal of Agricultural Science* 145 (5): 491–500.

Moseley, W. G., and L. C. Gray. 2008. *Hanging by a Thread: Cotton, Globalization and Poverty in Africa*. Athens, OH, US: Ohio University Press.

Muwanga-Zake, E. S. K. 2009. "First Panorama Report: Uganda." Institute of Statistics and Applied Economics, Makerere University. GCP/GLO/208/BMG–CountrySTAT for Sub-Saharan Africa. Accessed May 15, 2013. www.countrystat.org/country/UGA/contents/docs/Panorama_Report_I-UGA.pdf.

National Research Council. 2010. *The Impact of Genetically Engineered Crops on Farm Sustainability in the United States*. Washington, DC: National Academies Press.

Ndeffo Mbah, M. L., J. H. Forster, J. Wesseler, and C. A. Gilligan. 2010. "Economically Optimal Timing for Crop Disease Control under Uncertainty: An Options Approach." *Journal of the Royal Society Interface* 7 (51): 1421–1428.

Net-Map Toolbox: Influence Mapping of Social Networks. 2013. "About." Accessed February 13. http://netmap.wordpress.com/about/.

Norton, G. W., and D. Hautea. 2009. *Projected Impacts of Agricultural Biotechnologies for Fruits and Vegetables in the Philippines and Indonesia*. Los Baños, Laguna, Philippines: International Service for the Acquisition of Agri-biotech Applications and the South East Asian Ministers of Education Organization's Southeast Asian Regional Center for Graduate Study and Research in Agriculture.

Oehmke, J. F., and C. A. Wolf. 2004. "Why Is Monsanto Leaving Money on the Table? Monopoly Pricing and Technology Valuation Distributions with Heterogeneous Adopters." *Journal of Agricultural and Applied Economics* 36 (3): 705–718.

Ogwang, J., M. B. Sekamatte, and A. Tindyebwa. 2005. "Report on the Ground Situation of Organic Cotton Production in Selected Areas of the Lango Sub-Region." Mimeo, National Agricultural Research Organisation, US Agency for International Development / Uganda Agricultural Productivity Enhancement Program, Uganda Export Promotions Board, Kampala.

Pachico, D., J. K. Lynam, and P. G. Jones. 1987. "The Distribution of Benefits from Technical Change among Classes of Consumers and Producers: An ex ante Analysis of Beans in Brazil." *Research Policy* 16 (5): 279–285.

Palisade Corporation. 2012. @Risk (computer software). Ithaca, NY, US.

Pemsl, D., H. Waibel, and A. P. Gutierrez. 2005. "Why Do Some Bt-Cotton Farmers in China Continue to Use High Levels of Pesticides?" *International Journal of Agricultural Sustainability* 3 (1): 44–56.

Pemsl, D., H. Waibel, and J. Orphal. 2004. "A Methodology to Assess the Profitability of Bt-Cotton; Case Study Results from the State of Karnakata, India." *Crop Protection* 23 (12): 1249–1257.

Poulton, C., and D. Tschirley. 2009. "A Typology of African Cotton Sectors." In *Organization and Performance of Cotton Sectors in Africa, Learning from Reform Experience,* edited by D. Tschirley, C. Poulton, and P. Labaste, 45–54. Washington, DC: World Bank.

Poulton, C., P. Labaste, and D. Boughton. 2009. "Yields and Returns to Farmers." In *Organization and Performance of Cotton Sectors in Africa, Learning from Reform Experience*, edited by D. Tschirley, C. Poulton, and P. Labaste, 117–137. Washington, DC: World Bank.

Pray, C. E. 2010. "The Role of Socioeconomic Assessment in Decisions about the Approval of GMOs in India, China and South Africa." Paper presented at the World Congress of Environmental and Resource Economics, Montreal, June 29.

Pray, C. E., J. Huang, R. Hu, and S. Rozelle. 2002. "Five Years of Bt Cotton in China—The Benefits Continue." *Plant Journal* 31 (4): 423–430.

Pray, C. E., B. Ramaswami, J. Huang, R. Hu, P. Bengali, and H. Zhang. 2006. "Costs and Enforcement of Biosafety Regulations in India and China." *International Journal of Technology and Globalisation* 2 (1/2): 137–151.

Qaim, M. 2003. "Bt Cotton in India: Field Trial Results and Economic Projections." *World Development* 31 (12): 2115–2127.

———. 2009. "The Economics of Genetically Modified Crops." *Annual Review of Resource Economics* 1: 665–694.

Qaim, M., and A. de Janvry. 2005. "Bt Cotton and Pesticide Use in Argentina: Economic and Environmental Effects." *Environment and Development Economics* 10 (2): 179–200.

Qaim, M., and D. Zilberman. 2003. "Yield Effect of Genetically Modified Crops in Developing Countries." *Science* 299 (5608): 900–902.

Qaim, M., A. Subramanian, G. Naik, and D. Zilberman. 2006. "Adoption of Bt Cotton and Impact Variability: Insights from India." *Review of Agricultural Economics* 28 (1): 48–58.

Quemada, H. 2003. "Developing a Regulatory Package for Insect Tolerant Potatoes for African Farmers: Projected Data Requirements for Regulatory Approval in South Africa." Paper presented at the symposium Strengthening Biosafety Capacity for Development, Dikhololo, South Africa, June 9–11.

Raney, T. 2006. "Economic Impact of Transgenic Crops in Developing Countries." *Current Opinion in Biotechnology* 17 (2): 174–178.

Reardon, T., V. Kelly, E. Crawford, T. Jayne, K. Savadogo, and D. Clay. 1997. *Determinants of Farm Productivity in Africa: A Synthesis of Four Case Studies*. Department of Agricultural Economics, MSU Technical Paper 75. Michigan: Michigan State University.

Rose, R. N. 1980. "Supply Shifts and Research Benefits: A Comment." *American Journal of Agricultural Economics* 69 (1): 78–86.

Schiffer, E. 2007. *The Power Mapping Tool: A Method for the Empirical Research of Power Relations*. Discussion Paper 703. Washington, DC: International Food Policy Research Institute.

Schiffer, E., C. Narrod, and K. von Grebner. 2008. *The Role of Information Networks in Communicating and Responding to HPAI Outbreaks*. HPAI Research Brief 5. www.ifpri.org/ sites/default/files/publications/hpairb05.pdf.

Scobie, G. M., and R. T. Posada. 1978. "The Impact of Technical Change on Income Distribution: The Case of Rice in Colombia." *American Journal of Agricultural Economics* 60 (1): 85–92.

Serunjogi, L. K., P. Elobu, G. Epieru, V. A. O. Okoth, M. B. Sekamatte, J. P. Takan, and J. O. E. Oryokot. 2001. "Traditional Cash Crops: Cotton (*Gossypium* sp.)." In *Agriculture in Uganda, Volume II: Crops*, edited by J. K. Mukiibi, 322–375. Kampala: Fountain Publishers / CTA / National Agricultural Research Organisation.

Sexton, S., and D. Zilberman. 2011. *How Agricultural Biotechnology Boosts Food Supply and Accommodates Biofuels*. Working Paper 16699. Cambridge, MA: National Bureau of Economic Research.

Shankar, B., and C. Thirtle. 2005. "Pesticide Productivity and Transgenic Cotton Technology: The South African Smallholder Case." *Journal of Agricultural Economics* 56 (1): 97–116.

Smale, M., M. R. Bellon, and P. L. Pingali. 1998. "Farmers, Gene Banks, and Crop Breeding: Introduction and Overview." In *Farmers, Gene Banks, and Crop Breeding. Economic Analyses of Diversity in Wheat, Maize, and Rice,* edited by M. Smale. Mexico City and Boston: International Center for Maize and Wheat Improvement and Kluwer Academic.

Smale, M., A. Niane, and P. Zambrano. 2010. "Une Revue des Méthodes Appliquées dans l'Evaluation de l'Impact Economique des Plantes Transgéniques sur les Producteurs dans l'Agriculture Non-Industrialisée: La Première Décennie." *Economie Rurale* 315 (January–February): 60–81.

Smale, M., P. Zambrano, G. Gruère, J. Falck-Zepeda, I. Matuschke, D. Horna, L. Nagarajan, et al. 2009. *Measuring the Economic Impacts of Transgenic Crops in Developing Agriculture during the First Decade. Approaches, Findings, and Future Direction.* Food Policy Review 10. Washington, DC: International Food Policy Research Institute.

Sunding, D. L., and D. Zilberman. 2001. "The Agricultural Innovation Process: Research and Technology Adoption in a Changing Agricultural Sector." In *Handbook of Agricultural Economics,* volume 1, edited by B. L. Gardner and G. C. Rausser, 207–261. Amsterdam: Elsevier.

Taylor, A. 2006. *Overview of the Current State of Organic Agriculture in Kenya, Uganda and the United Republic of Tanzania and the Opportunities for Regional Harmonization.* New York and Geneva: United Nations.

Traxler, G., and S. Godoy-Avila. 2004. "Transgenic Cotton in Mexico." *AgBioForum* 7 (1–2): 57–62.

Traxler, G., S. Godoy-Ávila, J. Falck-Zepeda, and J. de J. Espinoza-Arellano. 2003. "Transgenic Cotton in Mexico: Economic and Environmental Impacts of the First Generation Biotechnologies." In *The Economic and Environmental Impacts of Agbiotech: A Global Perspective,* edited by N. Kalaitzandonakes, 183–202. New York: Kluwer Academic / Plenum.

Tripp, R. 2009. *Biotechnology and Agriculture Development: Transgenic Cotton, Rural Institutions and Resource-Poor Farmers.* London: Routledge.

Tschirley, D., C. Poulton, and P. Labaste. 2009. *Organization and Performance of Cotton Sectors in Africa.* Washington, DC: World Bank.

Tulip, A., and P. Ton. 2002. *Organic Cotton Study, Uganda Case Study: A Report for PAN UK's Pesticides and Poverty Project.* London: Pesticide Action Network UK.

UBOS (Uganda Bureau of Statistics). 2007. *Uganda National Household Survey 2005/2006: Report of the Agricultural Module.* Entebbe, Uganda: Uganda Bureau of Statistics. www.ubos.org/onlinefiles/uploads/ubos/pdf%20documents/2005UNHSAgriculturalModuleReport.pdf.

UEPB (Uganda Export Promotion Board). 2007. *Export Performance Watch.* Export Bulletin Edition 10. Kampala, Uganda.

Vitale, J. D., G. Vognan, M. Ouattarra, and O. Traore. 2010. "The Commercial Application of GMO Crops in Africa: Burkina Faso's Decade of Experience with Bt Cotton." *AgBioForum* 13 (4): 320–332.

Willert, H., and M. Yussefi. 2007. *The World of Organic Agriculture: Statistics and Emerging Trends 2007.* Bonn, Germany, and Frick, Switzerland: International Federation of Organic Agriculture Movements and Research Institute of Organic Agriculture.

Wilson, W. W., E. A. De Vuyst, R. D. Taylor, W. W. Koo, and B. L. Dahl. 2008. "Implications of Biotech Traits with Segregation Costs and Market Segments: The Case of Roundup Ready Wheat." *European Review of Agricultural Economics* 35 (1): 51–73.

Xu, N., M. Fok, L. Bai, and Z. Zhou. 2008. "Effectiveness and Chemical Pest Control of Bt-Cotton in the Yangtze River Valley." *Crop Protection* 27: 1269–1276.

Yanggen, D., V. Kelly, T. Reardon, and A. Naseem. 1998. *Incentives for Fertilizer Use in Sub-Saharan Africa: A Review of Empirical Evidence on Fertilizer Yield Response and Profitability.* MSU International Development Working Paper 70. East Lansing, MI, US: Michigan State University.

You, L., and J. Chamberlin. 2004. *Spatial Analysis of Sustainable Livelihood Enterprises of Uganda Cotton Production.* Environment and Production Technology Division Discussion Paper 121. Washington, DC: International Food Policy Research Institute.

Young, L. 2002. *Determining the Discount Rate for Government Projects.* Working Paper 02/21. Wellington, New Zealand: New Zealand Treasury.

Young, T. R. 2004. *Genetically Modified Organisms and Biosafety: A Background Paper for Decision-Makers and Others to Assist in Consideration of GMO Issues.* IUCN Policy and Global Change Series 1. Gland, Switzerland: International Union for the Conservation of Nature.

Zambrano, P., L. A. Fonseca, I. Cardona, and E. Magalhaes. 2009. "The Socioeconomic Impact of Transgenic Cotton in Colombia." In *Biotechnology and Agriculture Development: Transgenic Cotton, Rural Institution and Resource-Poor Farmers*, edited by R. Tripp, 168–199. London: Routledge.

Zhao, X., W. E. Griffiths, R. Griffith, and J. D. Mullen. 2000. "Probability Distributions for Economic Surplus Changes: The Case of Technical Change in the Australian Wool Industry." *Australian Journal of Agricultural and Resource Economics* 44 (1): 83–106.

Contributors

José Falck-Zepeda (j.falck-zepeda@cgiar.org) is a senior research fellow in the Environment and Production Technology Division of the International Food Policy Research Institute, Washington, DC, and leader of the polity team in IFPRI's Program for Biosafety Systems. His relevant publications include "Estimates and Implications of the Costs of Compliance with Biosafety Regulations in Developing Countries," *GM Crops and Food: Biotechnology and Agriculture in the Food Chain* 3 (1), coauthored with Jose Yorobe Jr., Bahagiawati Amir Husin, Abraham Manalo, Erna Lokollo, Godfrey Ramon, Patricia Zambrano, and Sutrisno; and "Socio-economic Considerations in Biosafety and Biotechnology Decision Making: The Cartagena Protocol and National Biosafety Frameworks," *Review of Policy Research* 28 (2), coauthored with Patricia Zambrano. Falck-Zepeda is the author of "Socio-economic Considerations, Article 26.1 of the Cartagena Protocol on Biosafety: What Are the Issues and What Is at Stake?" *AgBioForum* 12 (1), and coedited, with Melinda Smale, a special issue of *AgBioForum:* "Farmers and Researchers Discovering Biotech Crops: Experiences Measuring Economic Impacts among New Adopters," *AgBioForum* 15 (2).

Guillaume Gruère (ggruere@gmail.com) was a senior research fellow in the Environment and Production Technology Division of the International Food Policy Research Institute, Washington, DC. His relevant publications include "Refining Opportunity Cost Estimates of Not Adopting GM Cotton: An Application in Seven Sub-Saharan African

Countries," *Applied Economic Perspectives and Policy* 33 (2), coauthored with Antoine Bouët; "GM-Free Private Standards and Their Effects on Biosafety Decision-Making in Developing Countries," *Food Policy* 34 (5), coauthored with Debdatta Sengupta; "Will They Stay or Will They Go? The Political Influence of GM-Averse Importing Companies on Biosafety Decision Makers in Africa," *American Journal of Agricultural Economics* 94 (3), co-authored with Hiroyuki Takeshima.

Daniela Horna (jdhorna@fastmail.fm) is a freelance researcher in Washington, DC. She was a postdoctoral fellow in the Environment and Production Technology Division of the International Food Policy Research Institute, Washington, DC. Her relevant publications include "Farmer Willingness to Pay for Seed-Related Information: Rice Varieties in Nigeria and Benin," *Environment and Development Economics* 12 (6), coauthored with Melinda Smale and Matthias von Oppen; *Insecticide Use on Vegetables in Ghana: Would GM Seed Benefit Farmers?* IFPRI Discussion Paper 785, coauthored with Melinda Smale, Ramatu Al-Hassan, José Falck-Zepeda, and Samuel E. Timpo; and "Betting on Cotton: Potential Payoffs and Economic Risks of Adopting Transgenic Cotton in West Africa," *African Journal of Agricultural and Resource Economics* 2 (2), coauthored with José Falck-Zepeda and Melinda Smale.

John Komen (jce.komen@planet.nl) is assistant director and Africa coordi-nator for the Program for Biosafety Systems of the International Food Policy Research Institute, Washington, DC. His relevant publications include "The Emerging International Regulatory Framework for Biotechnology," *GM Crops and Food: Biotechnology and Agriculture in the Food Chain* 3 (1) and "Capacity Development for Agricultural Biotechnology and Biosafety Decision Making: Facilitating Implementation of Confined Field Trials in Uganda," *Proceedings of the International Conference on Agro-biotechnology, Biosafety and Seed Systems in Developing Countries,* coauthored with Theresa Sengooba. Komen also wrote "Capacity Building in Biosafety," coauthored with Karim M. Maredia, Cholani Weebadde, and Kakoli Ghosh; and "The Evolving International Regulatory Regime: Impact on Agricultural Development," coauthored with Silvia Salazar; both in *Environmental Safety of Genetically Engineered Crops.*

Miriam Kyotalimye (m.kyotalimye@asareca.org) is a programme assistant in the Policy Analysis and Advocacy Programme of the Association for Strengthening Agricultural Research in Eastern and Central Africa.

Theresa Sengooba (tsengooba@yahoo.com) is the Uganda coordinator for the Program for Biosafety Systems, Kampala. She also worked for Uganda's National Agricultural Research Organisation for 20 years. Her relevant publications include "Capacity Development for Agricultural Biotechnology and Biosafety Decision Making: Facilitating Implementation of Confined Field Trials in Uganda," *Proceedings of the International Conference on Agrobiotechnology, Biosafety and Seed Systems in Developing Countries*, coauthored with John Komen; "Biosafety Education Relevant to Genetically Engineered Crops for Academic and Non-academic Stakeholders in East Africa," *Electronic Journal of Biotechnology* 12 (1), coauthored with Rebecca Grumet et al.; and "Developing National Biosafety Systems," *Promoting Biosafety and Biosecurity within the Life Sciences: An International Workshop in East Africa*.

Patricia Zambrano (p.zambrano@cgiar.org) is a senior research analyst in the Environment and Production Technology Division of the International Food Policy Research Institute, Washington, DC. Her relevant publications include "Unweaving the Threads: The Experiences of Female Farmers with Biotech Cotton in Colombia," coauthored with Melinda Smale, Jorge H. Maldonado, and Sandra L. Mendoza, and "A Case of Resistance: Herbicide-Tolerant Soybeans in Bolivia," coauthored with Melinda Smale, Rodrigo Paz-Ybarnegaray, and Willy Fernández-Montaño, both in *AgBioForum* 15 (2); and "The Socio-economic Impact of Transgenic Cotton in Colombia," *Biotechnology and Agricultural Development: Transgenic Cotton, Rural Institutions and Resource-Poor Farmers,* coauthored with Luz Amparo Fonseca, Iván Cardona, and Eduardo Magalhaes.

Index

Page numbers for entries occurring in figures are followed by an *f;* those for entries in notes, by an *n;* and those for entries in tables, by a *t.*